IEE CONTROL ENGINEERING SERIES 58

Series Editors: Professor D. P. Atherton
Professor G. W. Irwin

Power-plant control and instrumentation

The control of boilers and HRSG systems

Other volumes in print in this series:

Power-plant control and instrumentation
The control of boilers and HRSG systems

David Lindsley

The Institution of Electrical Engineers

To my wife, Jo. Thanks for everything, especially your patience and eagle-eyed spotting of errors during the checking of this book.

Published by: The Institution of Electrical Engineers, London, United Kingdom

The Institution of Electrical Engineers,
Michael Faraday House,
Six Hills Way, Stevenage,
Herts. SG1 2AY, United Kingdom

British Library Cataloguing in Publication Data

A CIP catalogue record for this book
is available from the British Library

ISBN 0 85296 765 9

Printed in England by TJ International, Padstow, Cornwall

Contents

Preface

The aim of this book is to examine the control and instrumentation systems of the drum-type boilers and heat-recovery steam generators (HRSGs) that are used for the production of steam for turbines and industrial processes. My intention is to provide information to assist the designers, users and maintenance staff of such plants in understanding how these systems function.

The end product of the steam plant may be electricity that is exported to the grid, or it may be steam or hot water that is sent to a nearby process plant, factory or housing complex, but in each case the general principles of the control systems will be very similar. Nevertheless, the design of these systems is a specialised task, an art as much as a science, and in this introduction I aim to draw attention to the width and depth of knowledge that it demands.

The knowledge base one needs to design a control system for a boiler or HRSG is unusually wide. A power station is a complex entity, embracing a wide range of what I refer to as primary disciplines—physics, chemical engineering, thermodynamics, mechanical engineering and electrical engineering. It also involves control technology and computing—the secondary disciplines that combine two or more of the primary subjects.

The many machines in a power station operate together as an integrated, highly interdependent system. In practice, engineers involved in any area of power-station design, operation or maintenance must necessarily have their skills focused on just one of the primary disciplines, necessitating a high degree of training and experience in the relevant field: but to work effectively they will also need at least a basic understanding of all the others.

That may seem wide enough for anyone, but as soon as the focus narrows onto one area, the control and instrumentation systems (C&I) of power stations, it becomes apparent that the subject is even more demanding. Engineers working in this particular field must be proficient in the highly complex areas of control theory and computers (hardware, software or both—fast moving, ever changing subjects in their own rights), but in addition they should have at least a rudimentary understanding of the thermodynamics of steam generation and use, and of metallurgy, chemistry and mechanical design. In addition it may be necessary to understand how high-voltage heavy-current electrical systems work.

One of the problems faced by the industry is the need for control engineers who understand, and are competent in, the very demanding field of computer systems, as well as in the more traditional areas of engineering. But, whereas the quantity and variety of information required by the engineer has grown enormously over the past half-century, the period allocated to graduate training has not expanded beyond the same four or five years that I spent while I was being trained. And in my day computers were specialised things that one might, perhaps, study after graduating.

Beside being complicated, computer technology is beguiling. It is tempting, and intellectually satisfying, to sit at a keyboard tapping away and generating words, formulae or pictures on the screen. If a mistake is made, the thing simply doesn't work. At worst the system may 'crash', necessitating a reboot—a process that may, at worst, result in the loss of much carefully-constructed data. But that is all.

On the other hand, a computer controlling any power-station plant is in command of a huge process involving explosive mixtures of gases, steam at pressures and temperatures that become instantly lethal if anything goes wrong, and massive roaring turbines driving generators that produce megawatts of power at many tens of thousands of volts. A small mistake or lack of attention to detail in such a case can have consequences that will certainly be severe, probably very expensive and possibly tragic.

A power station is a complex thing, and its construction is a frantic, long drawn-out process involving many people, sometimes hundreds of them, working amid the difficulties of noise, dust and dirt, and extremes of temperature. Heavy items are craned or manhandled into position under a mess of cables and pipes, often with showers of sparks raining down from welding and cutting operations high above. An instrument lovingly installed on a pipe is all too often used as a foothold for a heavy-booted rigger reaching up to install an item on another pipe. Instrument cubicles are on occasion used as latrines by labourers who are caught short in the middle of a task. Many a control desk designed with an eye for artistic merit has come into violent contact with a massive steel girder being moved into position—and emerged the worse off!

When one moves away from Europe or North America things become even worse. (The expression 'debugging' takes on a new significance when one has to extract a large and aggressive cockroach from an I/O card rack.)

Across the world, cable trenches are dug and cables laid in them by electricians and labourers who have little or no understanding of electronics. Expert supervision has its limits. Even if much careful attention has been paid to defining earthing and screening requirements, all may be lost if the wrong type of cable gland is used at a single point, or if armoured cable is wrongly glanded. The fact that a malfunction has been caused by interference is difficult enough to determine. Trying to discover why and where the interference occurred in kilometres of cable trays and ducts snaking their way through a vast site is often an impossible task.

So why should anybody in their right minds *want* to work in a field that is often difficult, sometimes dangerous and always stressful? Everybody will have their own answer but, for me, the magic of this field is its huge scope and enormous challenges. In few other industries will one have to apply one's mind to technologies that are so wide-ranging and disparate as the thermodynamic processes of steam at 500°C and the operation of a high-speed data highway.

It is a varied, demanding and exciting field, and if in the course of explaining its complexities I can lure into the power-station C&I field a few people who might otherwise not have considered an engineering career, then I shall be pleased.

So what I have tried to do in this book is to provide an outline of the subject in a readable format. In doing this I have had to limit the depth of the coverage. I make no apologies for glossing over some topics and for simplifying some concepts. The experts in a particular field may well quibble with my explanations, but I would maintain that if the ideas work in practice, then that is an adequate starting point. It will always be possible to refine the detail later on.

I must try to explain how I have approached the practical aspects of boiler control and instrumentation. The rapid evolution of technology makes it dangerous to define any details of implementation (a photograph of today's state-of-the-art control room becomes very dated within only a few years!). For this reason I have tried to concentrate on the overall principles of each system, as I did with my earlier book on this subject, since the principles of three-element feed-water control, as implemented in a modern distributed control system, are virtually the same as those implemented in a 40-year-old pneumatic system fulfilling the same function. This time, however, in addition to information on system prin-ciples I have tried to provide practical information on transmitters, analysers, flame monitors, actuators and cabling. At the time of writing, developments in these areas seem to have reached something of a plateau, and I can only hope that the information I have provided will not become outdated too soon. In any event, I believe that the matters relating to transmitter and actuator installation and use will remain relevant well into the conceivable future.

Finally I would like to thank the many individuals and organisations who have made contributions to this book, either with direct contributions of diagrams and technical articles, or by the provision of information. In thanking the following, I do not wish to ignore all the others who have helped me with this work: Balfour Beatty Ltd., Croydon, Surrey; B.I.C.C. Components Ltd., Bristol Babcock Ltd., Kidderminster, Worcestershire; British Standards Institution, London; Copes-Vulcan Ltd., Winsford, Cheshire; Fireye Ltd., Slough, Berkshire; Howden Sirocco Ltd., Glasgow; Kvaerner Pulping Ltd., Gothenburg, Sweden; Measurement Technology Ltd., Luton, Bedfordshire; Mitsui Babcock Ltd., Crawley, West Sussex; National Power plc, Swindon, Wiltshire; Rosemount Engineering Company, Bognor Regis, West Sussex; Scottish Power plc, Glasgow; Solartron Ltd., Farnborough, Hampshire; and Watson Smith Ltd., Leeds.

Diagrammatic symbols

In spite of the existence of many recognised standards for instrumentation symbols [1], I have chosen to adopt a simple format which should be sufficient to explain the concepts that I want to communicate to the reader. These symbols would not be comprehensive enough to fully define the requirements within a full-scale control-system design task (for example, the controller symbol does not indicate whether or not auto/manual facilities are required, or the form that these should take). Nevertheless, I believe the diagrams will be easily understood by engineers.

In the context of the controllers themselves, it is worth mentioning that different terms are used in the USA and elsewhere to identify the same function. In particular, the plant parameter that is measured and fed to a controller is, in Europe, called the 'measured value,' while in the USA it is referred to as the 'process variable'. Also, when referring to controllers, the term 'reset' is often used in the USA instead of 'integral action'.

Abbreviations and terms used in this book

This book is addressed to people working across two very different disciplines: power-plant and control systems. Technical terms and abbreviations that are easily understood by professionals in one field can be bewildering to those who understand the other side, and so the following list is provided in an attempt to help readers understand the abbreviations and some of the terms that are used in the text and elsewhere in the industry.

1oo2	one-out-of-two voting
2oo2	two-out-of-two voting
2oo3	two-out-of-three voting
AC	alternating current
ASCII	American Standard Code for Information Interchange (a standard defining the codes used for communication between computers and between computers and their peripherals)
ADC	analogue-to-digital converter
A/M	auto/manual control facility
BSI	British Standards Institution
C&I	control and instrumentation
CCR	central control room
CCGT	combined-cycle gas-turbine plant
CE	European Community
CHP	combined heat and power (a type of plant that burns a fuel to produce electricity and steam that is used either to heat a nearby complex or by an industrial process)
CMR	continuous maximum rating (also MCR)
CPU	central processing unit
DAC	digital-to-analogue converter
DC	direct current
DCS	distributed control system

deterministic	A deterministic system is one in which events are dealt with in the exact order in which they occur. With some systems, events are dealt with by means which causes action to be taken in a sequence that is dictated by external constraints (such as polling). Such a system is not deterministic
DV	desired value
EEPROM	electrically erasable programmable read-only memory
EMC	electromagnetic compatibility
EMI	electromagnetic interference
FAT	factory acceptance test
FD	forced draught
FDS	functional design specification
FWR	feed-water regulator (control valve)
H/A	hand/automatic control facility
HP	high pressure (the definition is relative: on major central-station plant it is usually above 100 bar *g*)
HRSG	heat-recovery steam generator
IC	integrated circuit
ID	induced draught
IEC	International Electro-technical Commission
IEE	Institution of Electrical Engineers
IEEE	Institute of Electrical and Electronics Engineers
IP	intermediate pressure (a relative definition, see HP above)
ISO	International Standards Organisation
I/O	input and output
KKS	Kraftwerk Kennzeichensystem (power station designation system)
LAN	local-area network
LED	light emitting diode
load	the flow of steam, in kg/s, that is produced at any given time by the boiler or HRSG (sometime also the electrical load on the generator, in MW)
LP	low pressure (a relative definition, see HP above)
machine	turbo-generator or alternator
MCB	miniature circuit breaker
MCR	maximum continuous rating (also CMR), typically, the highest rate of steam flow that a boiler can produce for extended periods.
mill	a device (also known as a pulveriser) that is used to crush coal into fine powder before it is fed to the burners
MTBF	mean time between failure
MTTR	mean time to repair
MV	measured value (also known as 'process variable')
P&ID	piping and instrumentation diagram
PCB	printed circuit board
PF	pulverised fuel (coal)

PLC	programmable-logic controller
PSU	power supply unit
pulveriser	a device (also known as a mill) that is used to crush coal into fine powder before it is fed to the burners
PV	process variable (also known as 'measured value')
RAM	random-access memory
RDF	refuse-derived fuel
RFI	radio-frequency interference
ROM	read-only memory
RTC	real-time clock
SAT	site acceptance test
SCADA	supervisory, control and data-aquisition system
TUV	Technischer Uberwachungs Verein (German Technical Supervisory Association)
UART	universal asynchronous receiver/transmitter (an electronic device that controls communication with a peripheral)
UL	underwriters' laboratories
UPS	uninterruptible power supply
VDU	visual display unit (also termed a 'monitor' or 'screen')
WTE	waste-to-energy (a type of plant where waste is burned to produce electricity or heat for a district or industrial process)

Reference

1 ANSI/ISA-S5.1: Instrumentation symbols and identification. Instrument Society of America, Research Triangle Park, North Carolina, USA, 1992

Chapter 1

The basics of steam generation and use

1.1 Why an understanding of steam is needed

Steam power is fundamental to what is by far the largest sector of the electricity-generating industry and without it the face of contemporary society would be dramatically different from its present one. We would be forced to rely on hydro-electric power plant, windmills, batteries, solar cells and fuel cells, all of which are capable of producing only a fraction of the electricity we use.

Steam *is* important, and the safety and efficiency of its generation and use depend on the application of control and instrumentation, often simply referred to as C&I. The objective of this book is to provide a bridge between the discipline of power-plant process engineering and those of electronics, instrumentation and control engineering.

I shall start by outlining in this chapter the change of state of water to steam, followed by an overview of the basic principles of steam generation and use. This seemingly simple subject is extremely complex. This will necessarily be an overview: it does not pretend to be a detailed treatise and at times it will simplify matters and gloss over some details which may even cause the thermodynamicist or combustion physicist to shudder, but it should be understood that the aim is to provide the C&I engineer with enough understanding of the subject to deal safely with practical control-system design, operational and maintenance problems.

1.2 Boiling: the change of state from water to steam

When water is heated its temperature rises in a way that can be detected (for example by a thermometer). The heat gained in this way is called *sensible* because its effects can be sensed, but at some point the water starts to boil.

But here we need to look even deeper into the subject. Exactly what *is* meant by the expression 'boiling'? To study this we must consider the three basic states of matter: solids, liquids and gases. (A plasma, produced when the atoms in a gas become ionised, is often referred to as the fourth state of matter, but for most practical purposes it is sufficient to consider only the three basic states.) In its solid state, matter consists of many molecules tightly bound together by attractive forces between them. When the matter absorbs heat the energy levels of its molecules increase and the mean distance between the molecules increases. As more and more heat is applied these effects increase until the attractive force between the molecules is eventually overcome and the particles become capable of moving about independently of each other. This change of state from solid to liquid is commonly recognised as 'melting'.

As more heat is applied to the liquid, some of the molecules gain enough energy to escape from the surface, a process called evaporation (whereby a pool of liquid spilled on a surface will gradually disappear). What is happening during the process of evaporation is that some of the molecules are escaping at fairly low temperatures, but as the temperature rises these escapes occur more rapidly and at a certain point the liquid becomes very agitated, with large quantities of bubbles rising to the surface. It is at this time that the liquid is said to start 'boiling'. It is in the process of changing state to a vapour, which is a fluid in a gaseous state.

Let us consider a quantity of water that is contained in an open vessel. Here, the air that blankets the surface exerts a pressure on the surface of the fluid and, as the temperature of the water is raised, enough energy is eventually gained to overcome the blanketing effect of that pressure and the water starts to change its state into that of a vapour (steam). Further heat added at this stage will not cause any further detectable change in temperature: the energy added is used to change the state of the fluid. Its effect can no longer be sensed by a thermometer, but it is still there. For this reason it is called *latent*, rather then sensible, heat. The temperature at which this happens is called the 'boiling point'. At normal atmospheric pressure the boiling point of water is 100 °C.

If the pressure of the air blanket on top of the water were to be increased, more energy would have to be introduced to the water to enable it to break free. In other words, the temperature must be raised further to make it boil. To illustrate this point, if the pressure is increased by 10% above its normal atmospheric value, the temperature of the water must be raised to just above 102 °C before boiling occurs.

The steam emerging from the boiling liquid is said to be saturated and, for any given pressure, the temperature at which boiling occurs is called the *saturation temperature*.

The information relating to steam at any combination of temperature, pressure and other factors may be found in steam tables, which are nowadays available in software as well as in the more traditional paper form. These tables were originally published in 1915 by Hugh Longbourne Callendar (1863–1930), a British physicist. Because of advances in knowledge and measurement technology, and as a result of changing units of measurement, many different variants of steam tables are today in existence, but they all enable one to look up, for any pressure, the saturation temperature, the heat per unit mass of fluid, the specific volume etc.

Understanding steam and the steam tables is essential in many stages of the design of power-plant control systems. For example, if a designer needs to compensate a steam-flow measurement for changes in pressure, or to correct for density errors in a water-level measurement, reference to these tables is essential.

Another term relating to steam defines the quantity of liquid mixed in with the vapour. In the UK this is called the *dryness fraction* (in the USA the term used is *steam quality*). What this means is that if each kilogram of the mixture contains 0.9 kg of vapour and 0.1 kg of water, the dryness fraction is 0.9.

Steam becomes *superheated* when its temperature is raised above the saturation temperature corresponding to its pressure. This is achieved by collecting it from the vessel in which the boiling is occurring, leading it away from the liquid through a pipe, and then adding more heat to it. This process adds further energy to the fluid, which improves the efficiency of the conversion of heat to electricity.

As stated earlier, heat added once the water has started to boil does not cause any further detectable change in temperature. Instead it changes the state of the fluid. Once the steam has formed, heat added to it contributes to the total heat of the vapour. This is the sensible heat *plus* the latent heat *plus* the heat used in increasing the temperature of each kilogram of the fluid through the number of degrees of superheat to which it has been raised.

In a power plant, a major objective is the conversion of energy locked up in the input fuel into either usable heat or electricity. In the interests of economics and the environment it is important to obtain the highest possible level of efficiency in this conversion process. As we have already seen, the greatest efficiency is obtained by maximising the energy level of the steam at the point of delivery to the next stage of the process. When as much energy as possible has been abstracted from the steam, the fluid reverts to the form of cold water, which is then warmed and treated to remove any air which may have become entrained in it before it is finally returned to the boiler for re-use.

1.3 The nature of steam

As stated in the Preface, the boilers and steam-generators that are the subject of this book provide steam to users such as industrial plant, or housing and other complexes, or to drive turbines that are the prime movers for electrical generators. For the purposes of this book, such processes are grouped together under the generic name 'power plant'. In all these applications the steam is produced by applying heat to water until it boils, and before we embark on our study of power-plant C&I we must understand the mechanisms involved in this process and the nature of steam itself.

First, we must pause to consider some basic thermodynamic processes. Two of these are the Carnot and Rankine cycles, and although the C&I engineer may not make use of these directly, it is nevertheless useful to have a basic understanding of what they are how they operate.

1.3.1 The Carnot cycle

The primary function of a power plant is to convert into electricity the energy locked up in some form of fuel resource. In spite of many attempts, it has not proved possible to generate electricity in large quantities from the direct conversion of the energy contained in a fossil fuel (or even a nuclear fuel) without the use of a medium that acts as an intermediary. Solar cells and fuel cells may one day achieve this aim on a scale large enough to make an impact on fossil-fuel utilisation, but at present such plants are confined to small-scale applications. The water turbines of hydro-electric plants are capable of generating large quantities of electricity, but such plants are necessarily restricted to areas where they are plentiful supplies of water at heights sufficient for use by these machines.

Therefore, if one wishes to obtain large quantities of electricity from a fossil fuel or from a nuclear reaction it is necessary to first release the energy that is available within that resource and then to transfer it to a generator, and this process necessitates the use of a medium to convey the energy from source to destination. Furthermore, it is necessary to employ a medium that is readily available and which can be used with relative safety and efficiency. On plant Earth, water is, at least in general, a plentiful and cheap medium for effecting such transfers. With the development of technology during the twentieth century other possibilities have been considered, such as the use of mercury, but except for applications such as spacecraft where entirely new sets of limitations and conditions apply, none of these has reached active use, and steam is universally used in power stations.

The use of water and steam to provide motive power has a long history. In the first century AD Hero of Alexandria showed that steam leaving via nozzles attached to a heated container filled with water would cause the vessel to rotate, but in this simple machine (the aeolipile) the steam leaving the vessel was wasted and for continuous operation the process therefore necessitated continually replacing the water. With the nature of Hero's design, it was not a saimple task to refill the vessel while it was in operation, but even if a method had been found, using water in a one-way process like this necessitates the provision of endless supplies of that fluid. It was not until 1824 that a French engineer, Sadi Carnot, proposed a way to resolve this problem. He used a *cycle*, where the transfer medium is part of a closed loop and the medium is returned to its starting conditions after it has done the work required of it.

Carnot framed one of the two laws of thermodynamics. The first, Joule's law, had related mechanical energy to work: Carnot's law defined the temperature relations applying to the conversion of heat energy into mechanical energy. He saw that if this process were to be made reversible, heat could be converted into work and then extracted and re-used to make a closed loop. In his concept (Figure 1.1), a piston moves freely without encountering any friction inside a cylinder made of some perfectly insulating material. The piston is driven by a 'working fluid'. The cylinder has a head at one end that can be switched at will from being a perfect conductor to being a perfect insulator. Outside the cylinder are two bodies, one of which can deliver heat without its own temperature (T_1) falling, the other being a bottomless cold sink at a temperature (T_2) which is also constant.

The operation of the system is shown graphically in figure 1.2, which shows the pressure/volume relationship of the fluid in the cylinder over the whole cycle. As the process is a repeating cycle its operation can be studied from any convenient starting point, and we shall begin at the point A, where the cylinder head (at this time assumed to be a perfect conductor of heat), allows heat from the hot source to enter the cylinder. The result is that the medium begins to expand, and if it is allowed to expand freely,

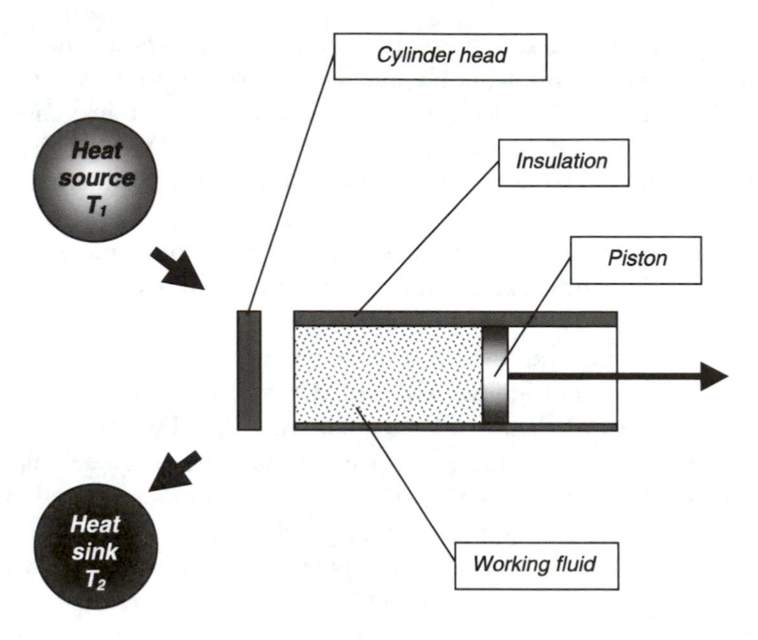

Figure 1.1 Carnot's heat engine

Boyle's law (which states that at any temperature the relationship between pressure and volume is constant) dictates that the temperature will not rise, but will stay at its initial temperature (T_1). This is called isothermal expansion.

When the pressure and volume of the medium have reached the values at point B, the cylinder head is switched from being a perfect conductor to being a perfect insulator and the medium allowed to continue its expansion with no heat being gained or lost. This is known as adiabatic expansion. When the pressure and volume of the medium reach the values at point C, the cylinder head is switched back to being a perfect conductor, but the external heat source is removed and replaced by the heat sink. The piston is driven towards the head, compressing the medium. Heat flows through the head to the heat sink and when the temperature of the medium reaches that of the heat sink (at point D), the cylinder head is once again switched to become a perfect insulator and the medium is compressed until it reaches its starting conditions of pressure and temperature. The cycle is then complete, having taken in and rejected heat while doing external work.

1.3.2 The Rankine cycle

The Carnot cycle postulates a cylinder with perfectly insulating walls and a head which can be switched at will from being a conductor to being

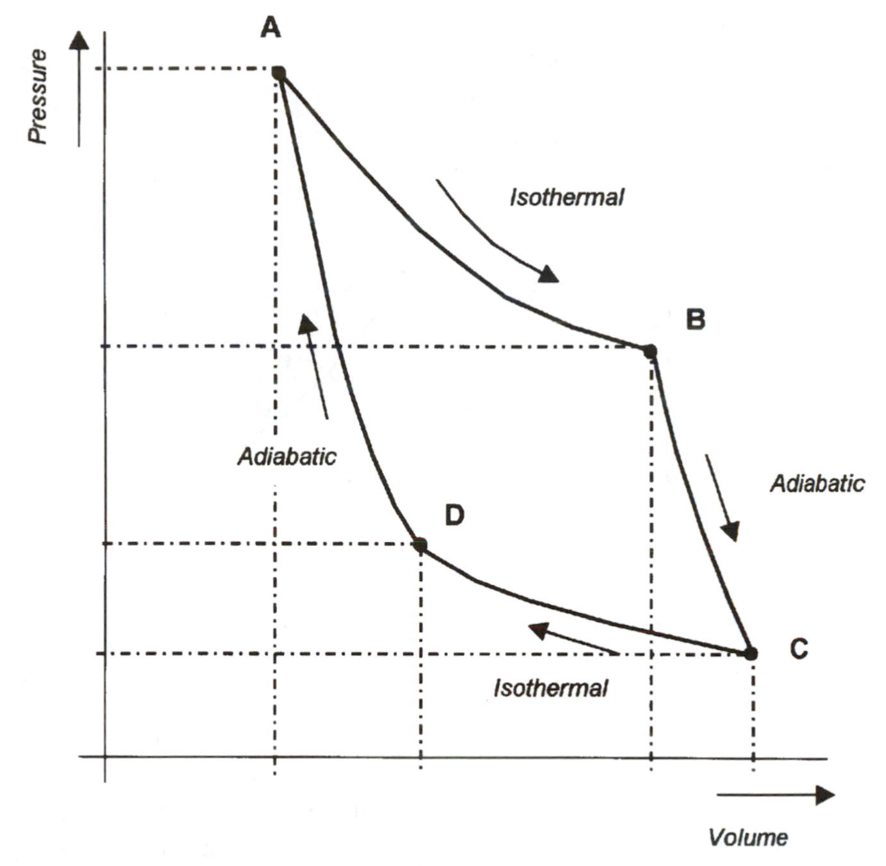

Figure 1.2 *The Carnot cycle*

an insulator. Even with modifications to enable it to operate in a world where such things are not obtainable, it would have probably remained a scientific concept with no practical application, had not a Scottish professor of engineering, William Rankine, proposed a modification to it at the beginning of the twentieth century [1]. The concepts that Rankine developed form the basis of all thermal power plants in use today. Even todays combined-cycle power plants use his cycle for one of the two phases of their operation.

Figure 1.3 illustrates the principle of the Rankine cycle. Starting at point A again, the source of heat is applied to expand the medium, this time at a constant pressure, to point B, after which adiabatic expansion is again made to occur until the medium reaches the conditions at point C. From here, the volume of the medium is reduced, at a constant pressure, until it reaches point D, when it is compressed back to its initial conditions.

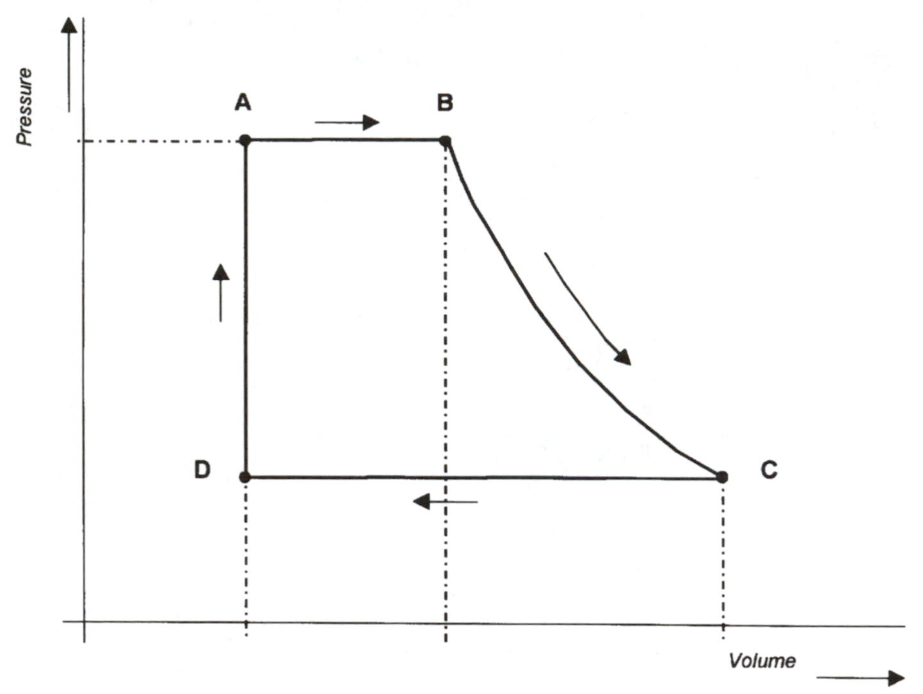

Figure 1.3 The Rankine cycle

All of this may seem of only theoretical interest, but it takes on a practical form in a thermal power plant, where water is compressed by pumps, then heated until it boils to produce steam which then expands (through a turbine or in some process) until it reverts to water. This operation is shown in Figure 1.4 which this time shows temperature plotted against a quantity called entropy for the processes within the boiler and turbine of a power plant. (Chapter 2 describes in detail the functions of the various items of plant.) Entropy is a measure of the portion of the energy in a system that is not available for doing work and it can be used to calculate heat transfer for a reversible process.

In the system shown in Figure 1.4, water is heated in feed heaters (A to B) using steam extracted from the turbine. Within the boiler itself, heat is used to further prewarm the water (in the economiser) before it enters the evaporative stages (C) where it boils. At D superheat is added until the conditions at E are reached at the turbine inlet. The steam expands in the turbine to the conditions at point F, after which it is condensed and returned to the feed heater. The energy in the steam leaving the boiler is converted to mechanical energy in the turbine, which then spins the generator to produce electricity.

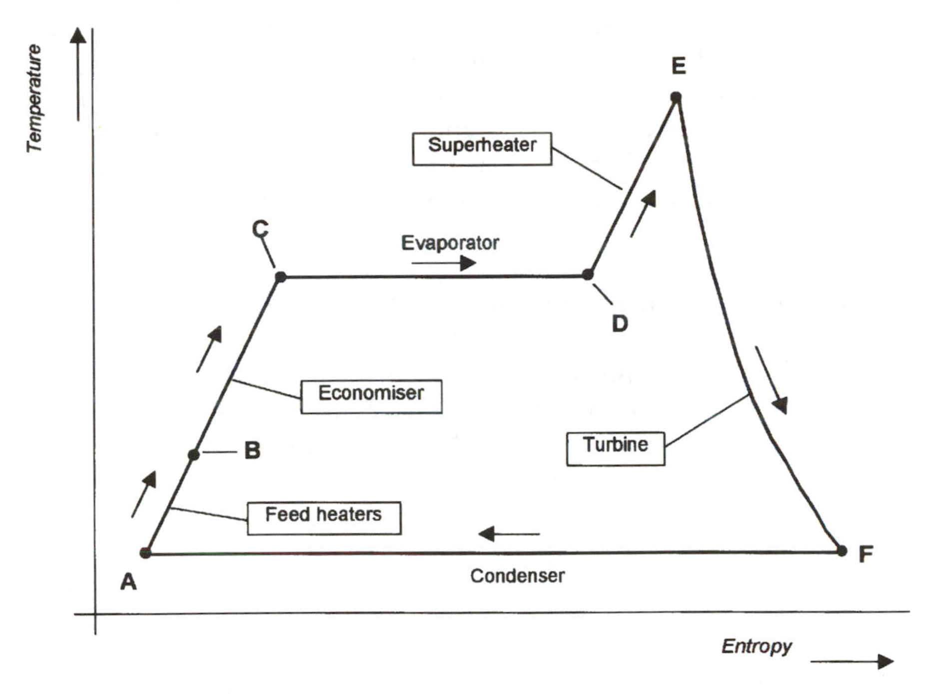

Figure 1.4 *The Rankine cycle in a steam-turbine power plant*

The diagram shows that the energy delivered to the turbine is maximised if point E is at the highest possible value and F is at the lowest possible value, and now we begin to see the importance of understanding these cycles if plant operation is to be understood and optimised. It explains why the temperature of the steam leaving the boiler is superheated and why the turbine condenser operates at very low pressures, which correspond with low temperatures.

1.4 Thermal efficiency

The efficiency of a power plant is the measure of its effectiveness in converting fuel into electrical energy or process heat. This factor sets the cost per unit of electricity or heat generated, and in a network of interconnected power stations it is this cost that determines the revenue that will be earned by the plant. Although several steps may be taken to reduce losses, some heat is inevitably lost in the flue gases and in the cooling water that leaves the condenser, and a realistic limit for the efficiency of such a plant is just over 40%. Although it has long been understood that, for every unit of money put into the operation of the plant, over half was being lost, very little could be done about this situation until developments in materials technology brought forward new opportunities.

One of the most dramatic power-plant developments of the second half of the twentieth century is the realisation that by employing one cycle in combination with another one, heat wasted in one could be use by the other to attain enhanced efficiency, this is the combined cycle.

1.5 The gas turbine and combined-cycle plants

The combined-cycle power station uses gas turbines to increase the efficiency of the power-generation process. Like many other machines that we assume to be products of the twentieth century, the gas turbine isn't that new. In fact, Leonardo da Vinci (1452–1519) sketched a machine for extracting mechanical energy from a gas stream. However, no practical implementation of such a machine was considered until the nineteenth century, when George Brayton proposed a cycle that used a combustion chamber exhausting to the atmosphere. In 1872 Germany's F. Stolze patented a machine that anticipated many features of a modern gas-turbine engine, although its performance was limited by the constraints of the materials available at the time.

Many other developments across Europe culminated in the development of an efficient gas turbine by Frank Whittle at the British Royal Aircraft Establishment (RAE) in the early 1930s. Subsequent developments at RAE led to viable axial-flow compressors, which could attain higher efficiencies than the centrifugal counterpart developed by Whittle.

All these gas turbines employed the Brayton cycle, whose pressure/volume characteristic is shown in Figure 1.5. Starting at point A in this cycle air is compressed isentropically (A–B) before being fed into a combustion chamber, where fuel is added and burned (B–C). The energy of the expanding air is then converted to mechanical work in a turbine (C–D). From C to D heat is rejected, and in a simple gas-turbine cycle this heat is lost to the atmosphere.

The rotation of the gas turbine can be used to drive a generator (via suitable reduction gearing) but, when used in a simple cycle with no heat recovery, the thermal efficiency of the gas turbine is poor, because of the heat lost to the atmosphere. The gases exhausted from the turbine are not only plentiful and hot (400–550 °C), but they also contain substantial amounts of oxygen (in combustion terms, the excess air level for the gas turbine is 200–300%). These factors point to the possibility of using the hot, oxygen-rich air in a steam-generating plant, whose steam output drives a turbine.

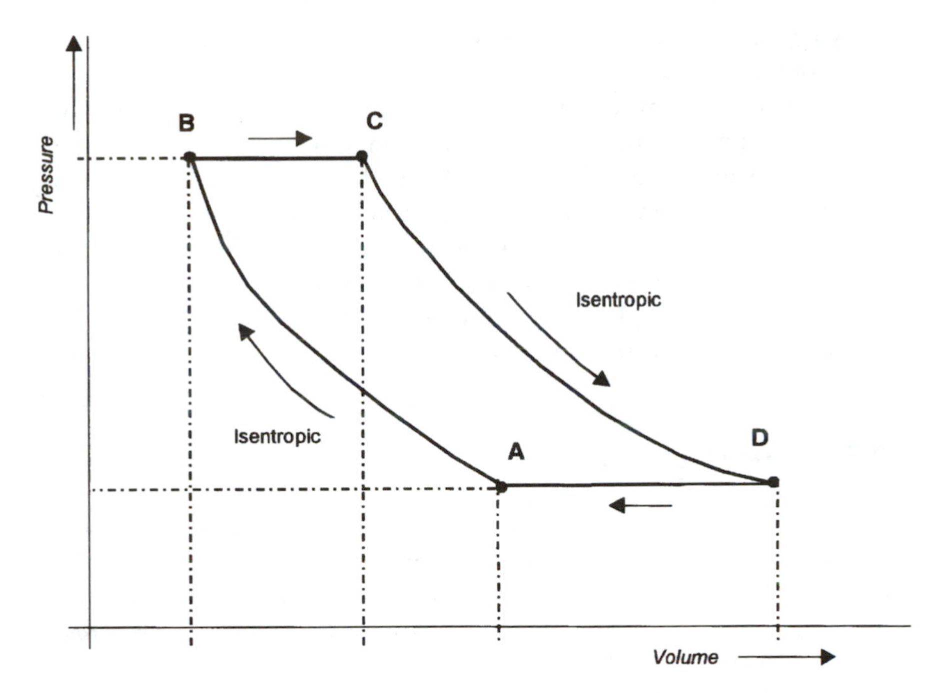

Figure 1.5 The Brayton cycle

The use of such otherwise wasted heat in a heat-recovery steam generator (HRSG) is the basis of the 'combined-cycle gas-turbine' (CCGT) plant which has been a major development of the past few decades. With the heat used to generate steam in this way, the whole plant becomes a binary unit employing the features of both the Rankine and the Brayton cycles to achieve efficiencies that are simply not possible with either cycle on its own. In fact, the addition of the HRSG yields a thermal efficiency that may be 50% higher than that of the gas turbine operating in simple-cycle mode.

Once again, there is nothing really new about this concept. From the moment when the gas turbine became a practical reality it was very obvious that the hot compressed air it exhausted contained huge amounts of heat. Therefore, the combined cycle was considered in some depth almost as soon as the gas turbine was released from the constraints of military applications. However, because of their use of gases at extremely high temperatures, early machines suffered from limited blade life and they were therefore used only in applications where no other source of power was readily available. With improvements in materials technology this difficulty has been overcome and, nowadays, combined-cycle plants employing gas turbines form the mainstream of modern power-station development.

But whether it is in a combined-cycle plant or a simple-cycle power station, our interest in this chapter is in steam and its use, and this vapour will now be examined in more detail. We shall see that what seems a fairly simple and commonplace thing is, in fact, quite complex.

In spite of its complexities it is important to tackle this subject in some depth, because the power-plant control and instrumentation engineer will need to deal with the physical parameters of steam through the various stages of designing or using a practical system.

1.6 Summary

In the above sections we have looked at the nature of steam and briefly explained how it is derived and used in various parts of the power station. We have also studied simple and combined cycles, and seen that the latter provide an opportunity of achieving higher efficiencies, thereby maximising the revenue earned by the plant.

In the following chapters we shall look at the plant in more detail, starting with the water and steam circuits and then moving on to discuss the combustion process. Once the plant is understood, the principles of its control systems can be better appreciated.

1.7 References

1 RANKINE, W.J.M.: 'A manual of the steam engine and other prime movers' (Griffin, London, 1908)

Chapter 2

The steam and water circuits

2.1 Steam generation and use

In a conventional thermal power plant, the heat used for steam generation may be obtained by burning a fossil fuel, or it may be derived from the exhaust of a gas turbine. In a nuclear plant the heat may be derived from the radioactive decay of a nuclear fuel. In this chapter we shall be examining the water and steam circuits of boilers and HRSGs, as well as the steam turbines and the plant that returns the condensed steam to the boiler.

In the type of plant being considered in this book, the water is contained in tubes lining the walls of a chamber which, in the case of a simple-cycle plant, is called the furnace or combustion chamber. In a combined-cycle plant the tubes form part of the HRSG. In either case, the application of the heat causes convection currents to form in the water contained in the tubes, causing it to rise up to a vessel called the *drum*, in which the steam is separated from the water. In some designs of plant the process of natural circulation is augmented by forced circulation, the water being pumped through the evaporative circuit rather than allowed to circulate by convection.

This book concentrates on plant where a drum is provided, but it is worth mentioning another type of plant where water passes from the liquid to the vapour stage without the use of such a separation vessel. Such 'once-through' boilers require feed-water and steam-temperature control philosophies that differ quite significantly from those described here, and they are outside the scope of this book.

Figure 2.1 shows a drum boiler in schematic form. Here, the steam generation occurs in banks of tubes that are exposed to the radiant heat of combustion. Of course, with HRSG plant no radiant energy is available

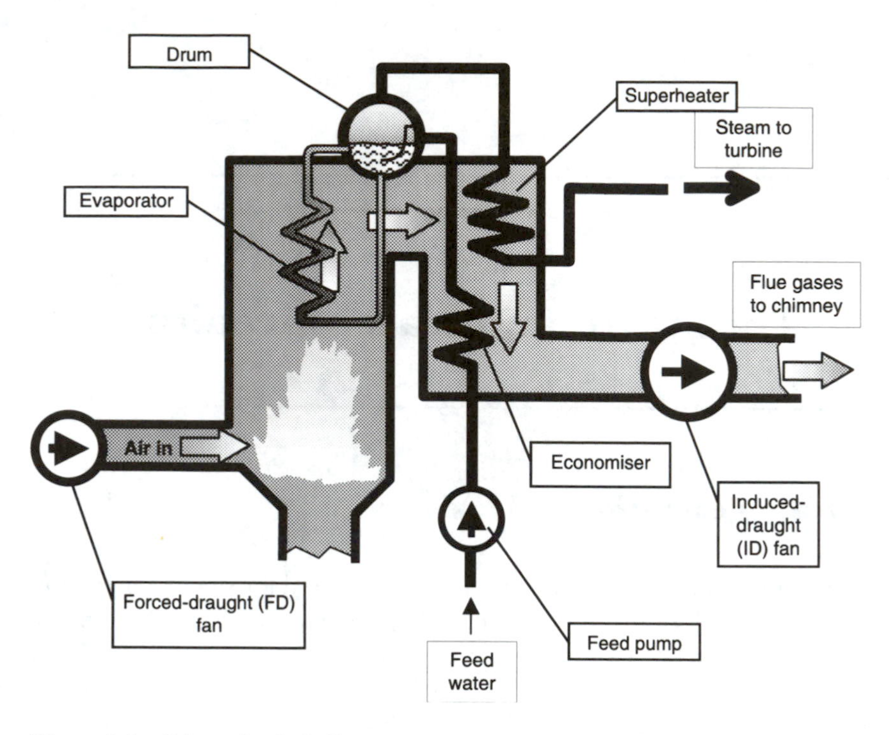

Figure 2.1 Schematic of a boiler

(since the combustion process occurs within the gas turbine itself) and the heat of the gas-turbine exhaust is transferred to the evaporator tubes by a mixture of convection and conduction. In this type of plant it is common to have two or more steam/water circuits (see Figure 2.6), each with its own steam drum, and in such plant each of these circuits is as described below.

The steam leaves the drum and enters a bank of tubes where more heat is taken from the gases and added to the steam, superheating it before it is fed to the turbine. In the diagram this part of the plant, the superheater, comprises a single bank of tubes but in many cases multiple stages of superheater tubes are suspended in the gas stream, each abstracting additional heat from the exhaust gases. In boilers (rather than HRSGs), some of these tube banks are exposed to the radiant heat of combustion and are therefore referred to as the radiant superheater. Others, the convection stages, are shielded from the radiant energy but extract heat from the hot gases of combustion.

After the flue gases have left the superheater they pass over a third set of tubes (called the economiser), where almost all of their remaining heat is extracted to prewarm the water before it enters the drum.

Finally the last of the heat in the gases is used to warm the air that is to be used in the process of burning the fuel. (This air heater is not shown in the diagram since it is part of the air and gas plant which is discussed in the next chapter.)

The major moving items of machinery shown in the diagram are the feed pump, which delivers water to the system, and the fan which provides the air needed for combustion of the fuel (in most plants each of these is duplicated). In a combined-cycle plant the place of the combustion-air fan and the fuel firing system is taken by the gas turbine exhaust.

Figure 2.1 shows only the major items associated with the boiler. In a power-generation station, the steam passes to a turbine after which it has to be condensed back to water, which necessitates the use of a heat exchanger to extract the last remaining vestiges of heat from the fluid and fully condense it into a liquid. Then, entrained air and gas has to be removed from the condensed fluid before it is returned to the boiler.

The major remaining plant items forming part of the steam/water cycle will now be briefly described and their operations explained.

2.2 The steam turbine

In plants using a turbine, the energy in the steam generated by the boiler is first converted to kinetic energy, then to mechanical rotation and finally to electrical energy. On leaving the turbine the fluid is fed to a condenser which completes the conversion back to water, which is then passed to further stages of processing before being fed to the feed pumps. In the following paragraphs, we shall examine this process (with the exception of the conversion to electrical energy in the alternator).

In the turbine, the steam is fed via nozzles onto successive rows of blades, of which alternate rows are fixed to the machine casing with the intermediate rows attached to a shaft (Figure 2.2). In this way the heat energy in the steam is converted first to kinetic energy as it enters the machine through nozzles, and then this kinetic energy is converted to mechanical work as it impinges onto the rotating blades. Further work is done by the reaction of the steam leaving these blades when it encounters another set of fixed blades, which in turn redirect it onto yet another set of rotating blades. As the steam travels through the machine in this way it continually expands, giving up some of its energy at each ring of blades. The moment of rotation applied to the shaft at any one ring of blades is the multiple of the force applied to the blades and mean distance of the force. Since each stage of rings abstracts energy from the steam, the force applied at the subsequent stage is less than it was at the preceding ring and,

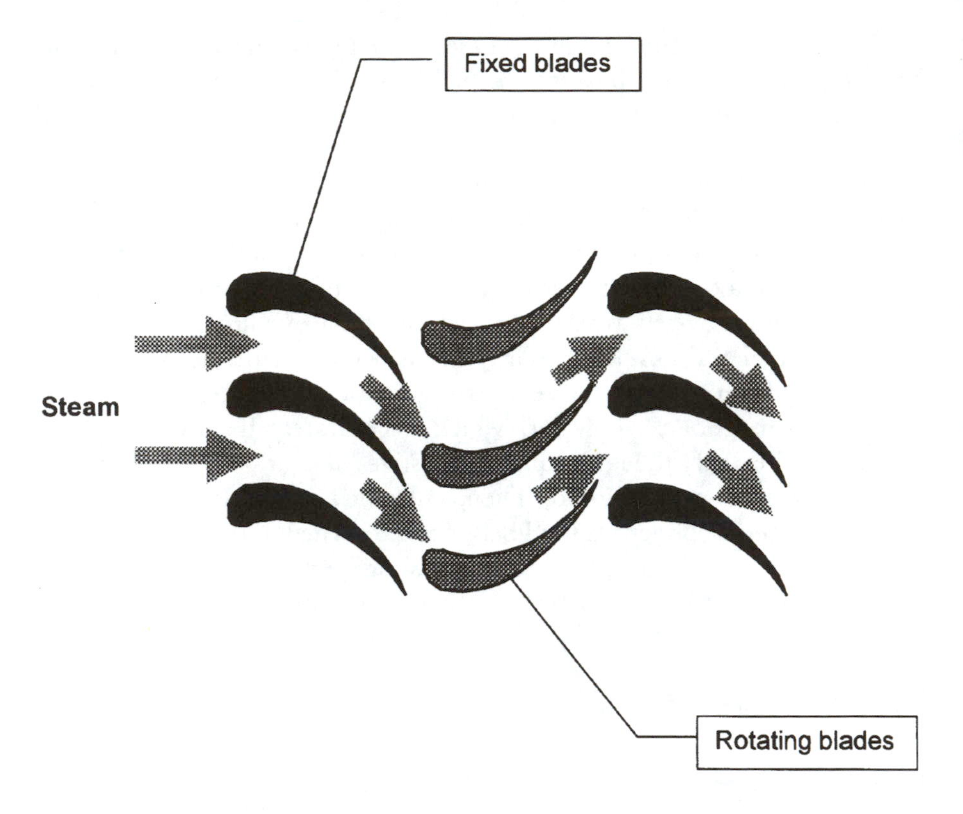

Figure 2.2 Turbine blading

therefore, to ensure that a constant moment is applied to the shaft at each stage, the length of the blades in all rings after the first is made longer than that of the preceding ring. This gives the turbine its characteristic tapering shape. The steam enters the machine at the set of blades with the smallest diameter and leaves it after the set of blades with the largest diameter. On the control diagrams presented in this book, this is indicated by the usual symbol for a turbine, a rhomboidal shape (Figure 2.3).

Turbines may consist of one or more stages, and in plant which uses reheating the steam exiting the high-pressure or intermediate stage of the machine (the HP or IP stage, respectively) is returned to the boiler for additional heat to be added to it in a bank of tubes called the reheater. The steam leaving this stage of the boiler enters the final stage of the machine, the low pressure (LP) stage. Because the energy available in the steam is now much less than it was at the HP stage, this part of the turbine is characterised by extremely long blades.

By the time it leaves the final stage of the turbine, the steam has exhausted almost all of the energy that was added to it in the steam generator, and it is therefore passed to a condenser where it is finally

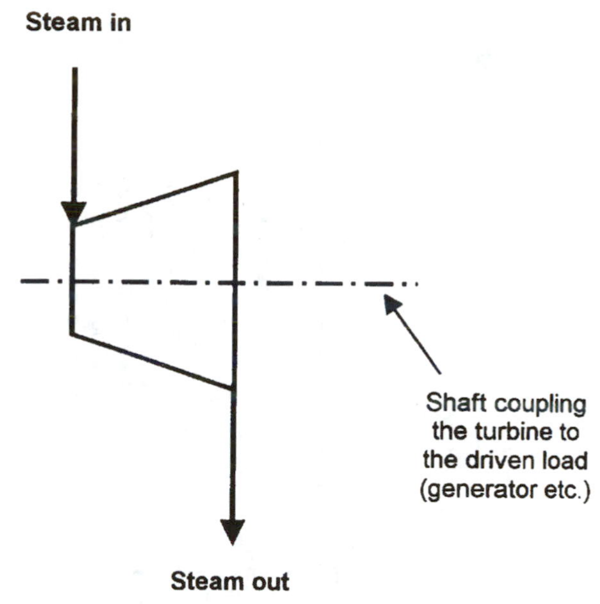

Figure 2.3 Symbolic representation of a turbine

cooled to convert it back to water which can be re-used in the cycle. The condenser comprises a heat exchanger through which cold water is circulated. A simplified representation of the complete circuit is shown in Figure 2.4.

The cooling water that is pumped through the condenser to abstract heat from the condensate may itself be flowing though a closed circuit. Alternatively, it may be drawn from a river or the sea to which it is then returned. In the latter cases, because of the heat received from the condenser, care must be taken to avoid undesirable heating of the river or sea in the vicinity of the discharge (or outfall).

In a closed circuit, the heat is released to the atmosphere in a cooling tower. Within these, the air that is used for cooling the water may circulate through the tower by natural convection, or it may be fan-assisted. It is usually desirable to minimise the formation of a plume since, as well as being very visible, such plumes can cause disturbance to the nearby environment by falling as a fine rain and possibly freezing on roads.

2.3 The condensate and feed-water system

Inside the plant, the steam and water system forms a closed loop, with the water leaving the condenser being fed back to the feed pumps for re-use in the boiler. However, certain other items of plant now become

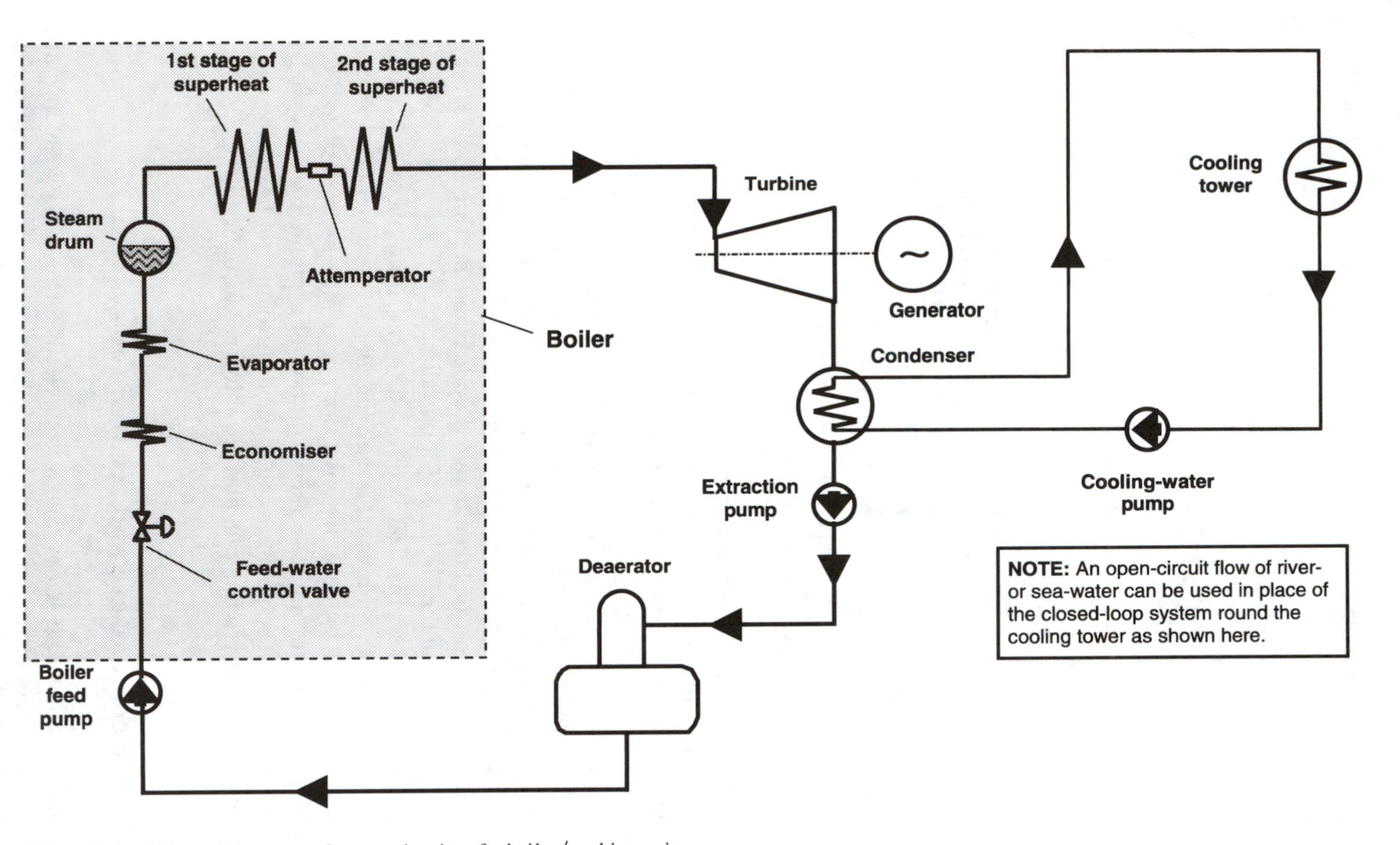

Figure 2.4 The main steam and water circuits of a boiler/turbine unit

involved, because the water leaving the condenser is cold and contains entrained air which must be removed.

Air becomes entrained in the water system at start-up (when the various vessels are initially empty), and it will appear during normal operation when it leaks in at those parts of the cycle which operate below atmospheric pressure, such as the condenser, extraction pumps and low-pressure feed heaters. Leakage can occur in these areas at flanges and at the sealing glands of the rotating shafts of pumps. Air entrainment is aided by two facts: one is that cold water can hold greater amounts of oxygen (and other dissolved gases) than can warm water; and the other is that the low-pressure parts of the cycle must necessarily correspond with the low-temperature phases.

The presence of residual oxygen in the feed-water supply of a boiler or HRSG is highly undesirable, because it will cause corrosion of the boiler pipework (particularly at welds, cold-worked sections and surface discontinuities), greatly reducing the serviceable life-span of the plant. For this reason great attention must be paid to its removal.

Removal of dissolved oxygen is performed in several ways, and an important contributor to this process is the deaerator which is shown in Figure 2.4, located between the condenser extraction pump and the boiler feed-water pump.

2.3.1 The deaerator

The deaerator removes dissolved gases by vigorously boiling the water and agitating it, a process referred to as 'stripping'. One type of deaerator is shown in Figure 2.5. In this, the water entering at the top is mixed with steam which is rising upwards. The steam, taken directly from the boiler or from an extraction point on the turbine, heats a stack of metal trays and as the water cascades down past these it mixes with the steam and becomes agitated, releasing the entrained gases. The steam pressurises the deaerator and its contents so that the dissolved gases are vented to the atmosphere.

Minimising corrosion requires the feed-water oxygen concentration to be maintained below 0.005 ppm or less and although the deaerator provides an effective method of removing the bulk of entrained gases it cannot reduce the concentration below about 0.007 ppm. For this reason, scavenging chemicals are added to remove the last traces of oxygen.

2.3.1.1 Chemical dosing

Volatile oxygen scavengers such as hydrazine (N_2H_4) and sodium sulphite (Na_2SO_3) have been used for oxygen removal (although

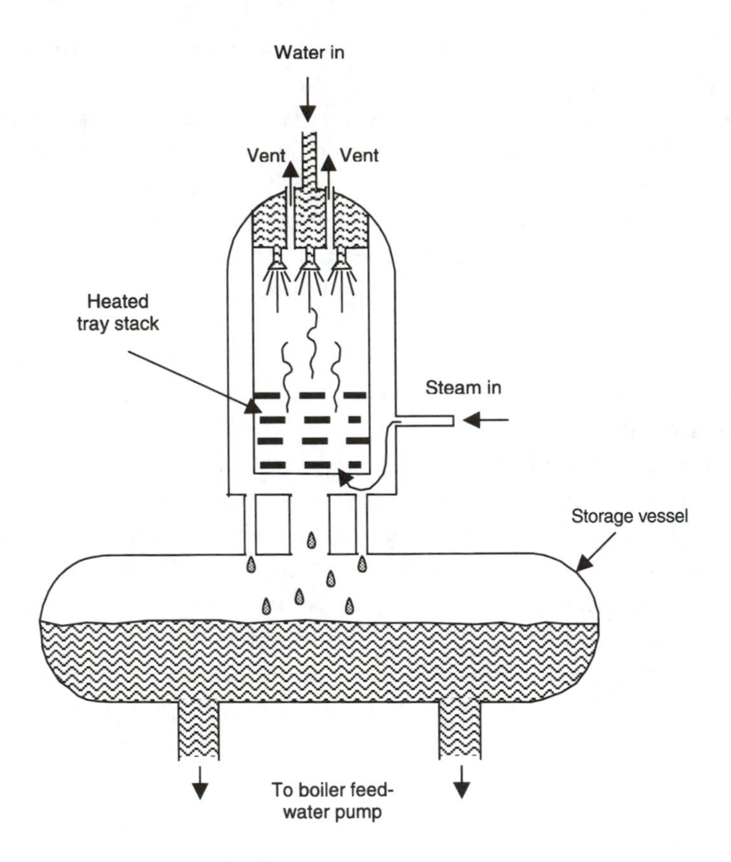

Figure 2.5 Principle of a deaerator

hydrazine is now suspected of being carcinogenic). Whatever their form, the chemical scavengers are added in a concentrated form and it is necessary to flush the injection pipes continually or on a periodic basis to prevent plugging. Similarly, *blowdown*, a process of bleeding water to drains or a special vessel, is used to continually or periodically remove a portion of the water from the boiler, with automatic or manual chemical sampling being used to ensure that the correct concentration is maintained in the boiler water.

From a control and instrumentation viewpoint, the above chemical dosing operations are highly specialised and are therefore usually performed by equipment that is supplied as part of a water-treatment plant package. The control system (often based on a programmable-logic control system (PLC)) will generate data and alarm signals for connection to the main plant computer-control system (frequently referred to as the distributed control system (DCS).)

After the water has been deaerated and treated, it is fed to feed pumps which deliver it back to the boiler at high pressure.

2.4 The feed pumps and valves

The feed pumps deliver water to the boiler at high pressure, and the flow into the system is controlled by one or more feed-regulating valves. The feed pumps are generally driven by electric motors, but small steam turbines are also used (although, clearly, these cannot be used at start-up unless a separate source of steam is available for their operation).

The pressure/flow characteristic of pumps and the various configurations that are available are discussed in Chapter 6 but it should be noted here that with any pump the pressure tends to fall as the throughput rises. On the other hand, due to the effect of friction, the resistance offered by the boiler system to the flow of water increases as the flow rate increases. (The system resistance is the minimum pressure that is required to force water into the boiler.) Therefore the pressure drop across the valve will be highest at low flows.

It is wasteful to operate with a pressure drop that is significantly above that at which effective control can be maintained, both because this entails an energy loss and also because erosion of valve internals increases with high pressure-drops. With fixed-speed pumps there is nothing that can be done about this, but an improvement can be made if variable-speed pumps are used. These are more expensive than their fixed-speed counterparts, but the increase in cost tends to be offset by the operational cost savings that can be achieved (due to more efficient operation and reduced wear on the valve). Such savings are increased if the plant operates for prolonged periods at low throughputs and are most apparent with the larger boilers.

From the control engineer's viewpoint, variable-speed pumps are an attractive option because they enable the control-system dynamics to be linearised over a wide range of flows, leading to improved controllability. However, the decision on their use will generally be made by mechanical and process engineers, and will be based purely on economic grounds.

2.5 The water and steam circuits of HRSG plant

In the combined-cycle plant the task of boiling the feed water and superheating the steam so produced is achieved by using the considerable heat

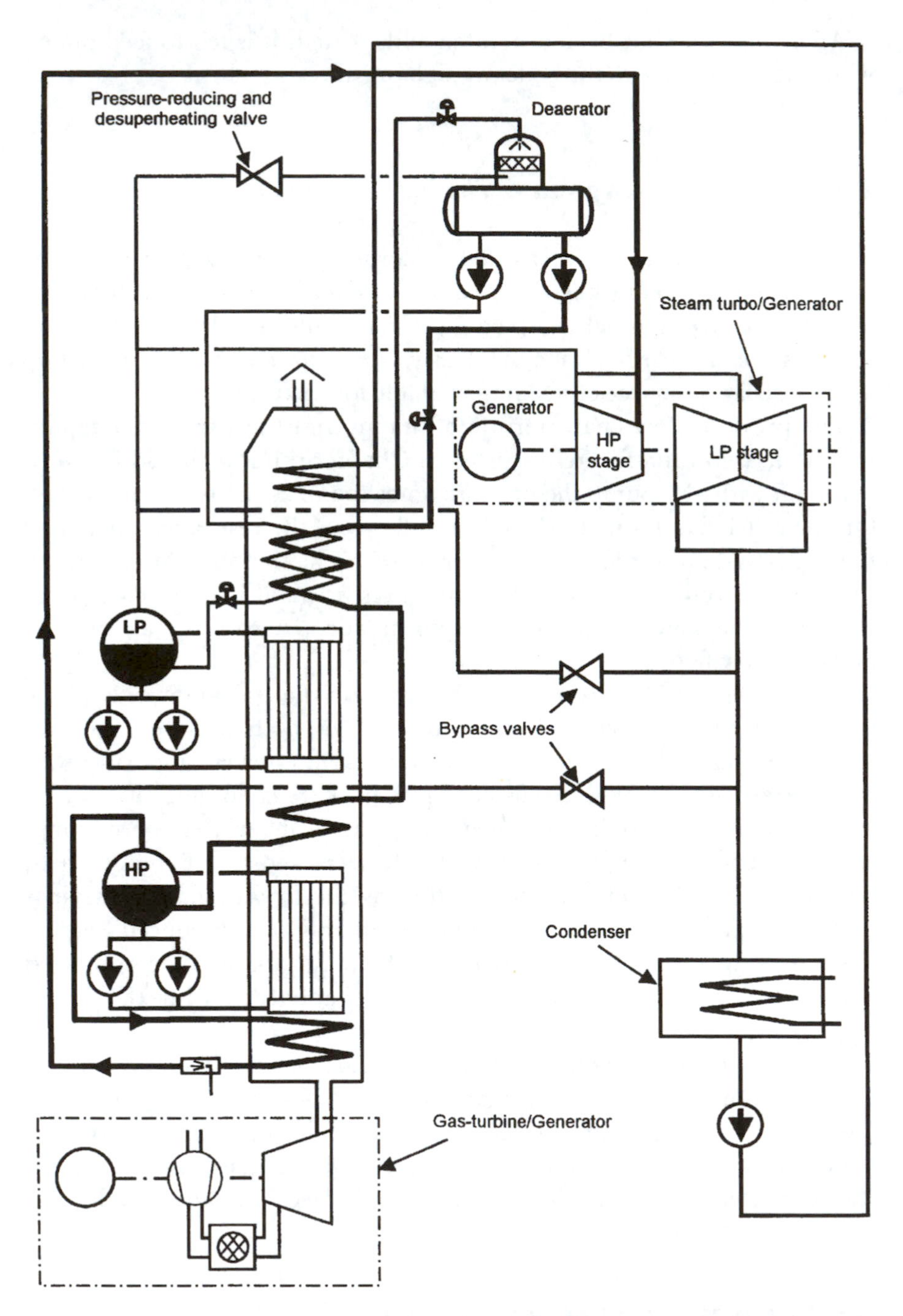

Figure 2.6 Gas turbines in a combined-cycle system

content of the exhaust from a gas turbine, sometimes with and sometimes without supplementary firing.

The variety of plant arrangements in use is very wide and although the following description relates to only one configuration, it should enable the general nature of these systems to be understood.

In some plants the gas and steam turbines and the generator are on the same shaft, others have separate generators for the gas and steam turbines. The installation shown in Figure 2.6 is of the latter variety, and the diagram shows just one gas turbine and HRSG from several at this particular plant.

Starting at the condenser outlet, the circuit can be traced through the extraction pump and via the economiser to the deaerator. From here two circuits are formed, one feeding the LP section, the other the HP section. These systems are of the forced-circulation type and are quite similar to each other in layout, but the steam leaving the HP side passes to a superheater bank which is positioned to receive the hottest part of the exhaust from the gas turbine. The superheated steam goes to the HP stage of the steam turbine and the steam leaving this stage goes to the LP stage. Saturated steam from the LP section of the HRSG also enters the turbine at this point. Bypass valves are employed during start-up and shut-down and enable the plant to operate with only the gas turbine in service, under which condition the steam from the HP and LP stages is bypassed to the condenser

2.6 Summary

So far, we have studied the nature of steam, and the plant and auxiliaries that are employed in the process of generating and using the fluid. Now we need to understand the mechanisms involved in obtaining the heat that is required to generate the steam. This process involves the fuel, air and flue-gas circuits of the plant, and all the major equipment required for clean and efficient operation.

Chapter 3 describes the combustion chamber (or furnace) and the plant and firing arrangements that are employed in burning a variety of fuels. In addition, the chapter outlines how the air required for combustion is obtained, warmed and distributed, and discusses the characteristics and limitations of the plant involved in this process.

Chapter 3

The fuel, air and flue-gas circuits

Having looked at the steam and water circuits of boilers and HRSGs, we now move on to examine the plant which is involved in the combustion of fuel in boilers.

The heat used for generating the steam is obtained by burning fuel in a furnace, or combustion chamber, but to do this requires the provision of air which is provided by a forced-draught (FD) fan (in larger boilers, two such fans are provided). After the fuel has been burned, the hot products of combustion are extracted from the furnace by another fan, the induced-draught (ID) fan, and fed to the chimney. Again, two ID fans are provided on larger boilers.

In this chapter we shall examine not only the burners or other equipment used to burn the fuel but also the fans and air heaters. Finally we shall briefly examine how gas turbines are used in combined-cycle plant.

3.1 The furnace

In boiler plant the heat used for boiling the water is obtained by burning a fossil fuel (unlike the HRSG, where the heat is delivered by the exhaust of a gas turbine). This process of combustion is carried out in the furnace, and comprises a chemical reaction between the combustible material and oxygen. If insufficient oxygen is available some of the combustibles will not burn, which is clearly inefficient and polluting. On the other hand, the provision of too much oxygen leads to inefficient operation and to corrosion and undesirable emissions from the stack due to the combination of the surplus oxygen with other components of the flue gases.

The oxygen for combustion is provided in air, which contains around 21% of the gas. However, air also contains around 77% nitrogen, and the combustion process results in the production of nitrogen dioxide (NO_2) and nitric oxide (NO). These gases (plus nitrous oxide, N_2O) are collectively called nitrogen oxides, or NO_x for short, and because they are often blamed for various detrimental effects on the environment a high level of attention must be given to minimising their production.

Unfortunately, high combustion efficiencies invariably correspond with the production of high levels of nitrogen oxides, and therefore NO_x reduction involves careful design of the burners so as to yield adequate combustion efficiency with minimal smoke and carbon monoxide generation.

3.1.1 Firing arrangements

The combustion of oil, gas or pulverised coal is performed in burners. These may be arranged on one wall of the combustion chamber (which is therefore called 'front-fired'), or on facing walls ('opposed fired') or at the corners of it ('corner-fired' or 'tangential'), and the characteristics of combustion will be very different in each case. The burners may be provided with individually controlled fuel and air supplies, or common control may be applied for all the burners, or they may be operated in groups, each group having dedicated and separately controlled supplies of fuel and air. Combustion of raw coal or other solid fuels such as municipal waste, clinical waste or refuse-derived fuels is often carried out in fluidised beds, or on stokers consisting of moving grates or platforms.

The methods of controlling these various arrangements are very different. With front-fired or opposed-fired boilers the temperature of the flue gases and the resulting heat transfer to the various banks of superheater tubes is adjusted by bringing burners into service or taking them out of service, and this may be done individually or in banks. This method provides a step-function type of control and fine adjustment of steam temperature is provided by spray-water attemporation.

Corner-fired (tangential) boilers are arranged in such a way that the burning fuel circulates around the furnace, forming a large swirling ball of burning fuel at the centre. With this type of boiler the manufacturers usually employ tilting mechanisms to direct the fireball to a higher or lower position within the furnace, and this has a significant effect on the temperature of the various banks of superheater tubes, and therefore on steam temperature. The maximum degree of tilt that is available within the basic design is typically $\pm 30°$, although the degree of movement employed in practice is usually restricted during commissioning.

The downside of tilting is that burners—with their fuel and air supplies, igniters, flame monitors etc.—are complex things, and tilting them requires very careful engineering if it is to be successful. Also, the tilting mechanisms must be rigorously maintained if they are to continue to operate effectively over any length of time.

The control systems that regulate burner tilting mechanisms must ensure that exactly the same degree of tilt is applied to the burners at all four corners of the furnace, since any misalignment will cause the fireball to circulate helically rather than as required.

3.2 The air and gas circuits

The combustion process requires the provision of fuel and air in the correct ratio to each other. This is known as the stoichiometic ratio, and under this condition enough air is provided to ensure complete combustion of all the fuel, with no surplus or deficit. However, this is a theoretical ideal, and practical considerations may necessitate operating at a fuel/air ratio that is different from the stoichiometric value. In addition, it must be understood that the efficiency of the combustion process will also be affected by the temperature of the air provided.

In the following sections we shall see how air is delivered to the furnace at the right conditions of flow and temperature, starting with the auxiliary plant that warms the air and moving on to the types of fan employed in the draught plant.

3.2.1 The air heater

In a simple-cycle plant, air is delivered to the boiler by one or more FD fans and the products of combustion are extracted from it by ID fans. Figure 3.1 shows this plant in a simplified form, and illustrates how the heat remaining in the exhaust gases leaving the furnace is used to warm the air being fed to the combustion chamber. This function is achieved in an air heater, which can be either regenerative, where an intermediate medium is used to transfer the heat from the exhaust gases to the incoming air, or recuperative, where a direct heat transfer is used across a dividing partition.

One variety of regenerative air heater is the Ljungström type, where metal plates mounted on a rotating frame are passed through the hot gases and then to the incoming air.

From a control engineer's point of view, an important consideration is the efficient combustion of the fuel, and here it is necessary to consider the

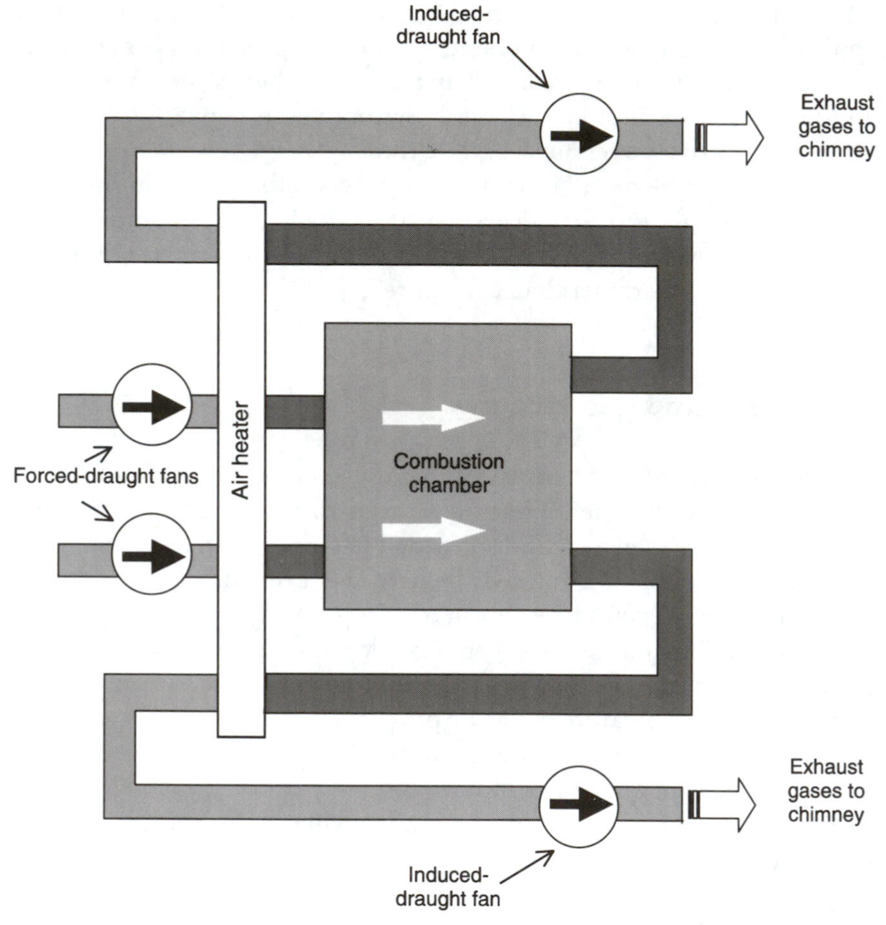

Figure 3.1 Draught-plant arrangement

losses and leakages that occur in an air heater. Figure 3.2 shows how various leakages occur in a typical air heater: across the circumferential, radial and axial seals, as well as at the hub. These leakages are minimised when the plant is first constructed, but become greater as wear occurs during prolonged usage. When the sheer physical size of the air heater is considered (Figure 3.3) it will be appreciated that these leakages can become significant.

3.2.2 Types of fan

In addition to the FD and ID fans mentioned above, another application for large fans in a power-station boiler is where it is necessary to overcome the resistance presented by plant in the path of the flue gases to the stack.

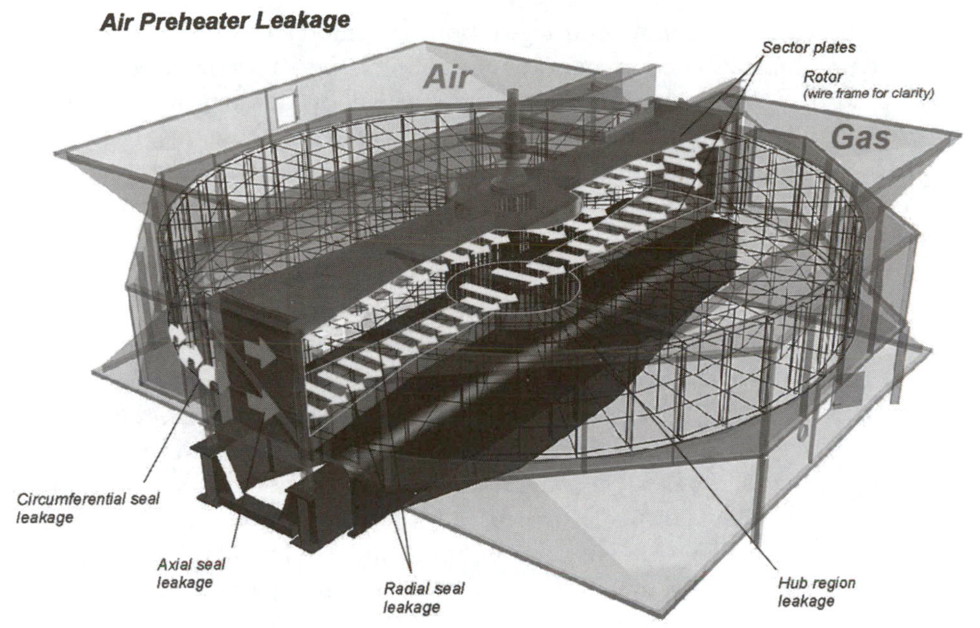

Figure 3.2 *Air heater leakage*
© Howden Sirocco Ltd. Reproduced with permission

Figure 3.3 *An air heater being lifted into position*
© Howden Sirocco Ltd. Reproduced with permission

In some cases, environmental legislation has enforced the fitting of flue-gas desulphurisation equipment to an existing boiler. This involves the use of absorbers and/or bag filters, plus the attendant ducting, all of which present additional resistance to the flow of gases. In this case this resistance was not anticipated when the plant was originally designed, so it is necessary to fit additional fans to overcome the draught losses. These are called 'booster fans'.

Whatever their function, as far as the fans themselves are concerned, two types are found in power-station draught applications: centrifugal (Figure 3.4) and axial-flow (Figure 3.5). In the former, the blades are set radially on the drive shaft with the air or flue gas directed to the centre and driven outwards by centrifugal force. With axial-flow fans, the air or gas is drawn along the line of the shaft by the screw action of the blades. Whereas the blades of a centrifugal fan are fixed rigidly to the shaft, the pitch of axial-flow fan blades can be adjusted. This provides an efficient means of controlling the fan's throughput, but requires careful design of the associated control system because of a phenomenon known as 'stall', which will now be described.

Figure 3.4 Centrifugal fan
© Howden Sirocco Ltd. Reproduced with permission

Figure 3.5 Axial-flow fan
© Howden Sirocco Ltd. Reproduced with permission

3.2.2.1 The stall condition

The angular relationship between the air flow impinging on the blade of a fan and the blade itself is known as the 'angle of attack'. In an axial-flow fan, when this angle exceeds a certain limit, the air flow over the blade separates from the surface and centrifugal force then throws the air outwards, towards the rim of the blades. This action causes a build-up of pressure at the blade tip, and this pressure increases until it can be relieved at the clearance between the tip and the casing. Under this condition the operation of the fan becomes unstable, vibration sets in and the flow starts to oscillate. The risk of stall increases if a fan is oversized or if the system resistance increases excessively.

For each setting of the blades there is a point on the fan characteristic beyond which stall will occur. If these points are linked, a 'stall line' is generated (Figure 3.6) and if this is built into the plant control system (DCS) it can be used to warn the operator that the condition is imminent and then to actively shift operation away from the danger region. The actual stall-line data for a given machine should be provided by the fan manufacturer.

3.2.2.2 Centrifugal-fan surge

The stall condition affects only axial-flow fans. However, centrifugal fans are subject to another form of instability. If they are operated near the peak of their pressure/flow curve a small movement either way can cause the pressure to increase or decrease unpredictably. The point at which this phenomenon occurs is known as the 'surge limit' and it is the minimum flow at which the fan operation is stable.

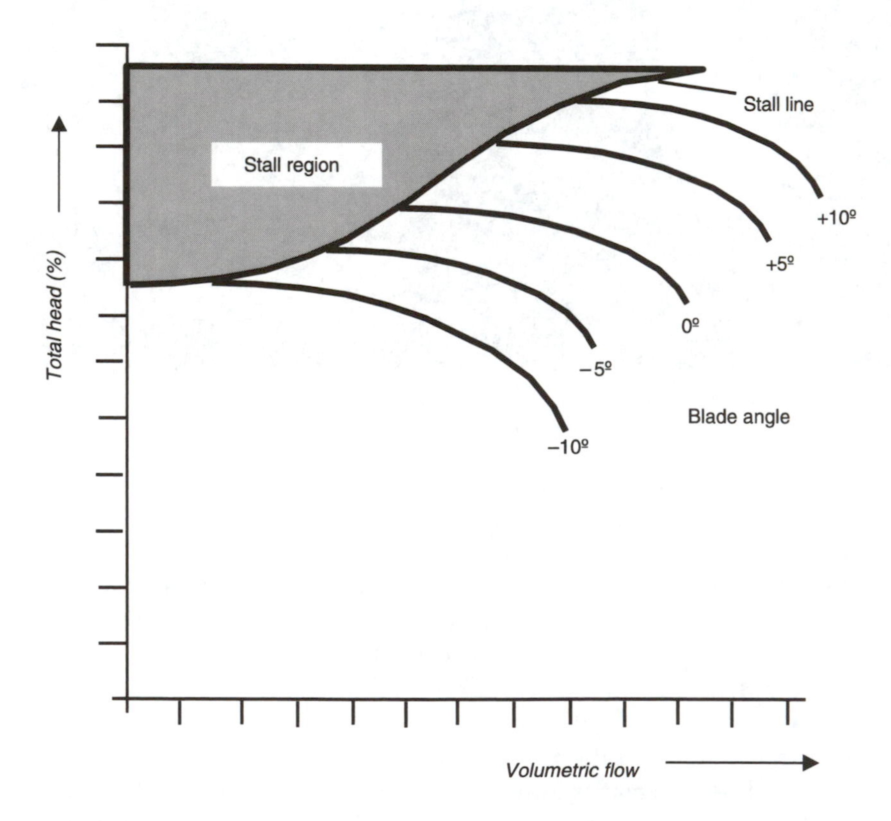

Figure 3.6 The stall line of an axial-flow fan

The system designer needs to be aware of the risk of surge occurring, since it may be necessary to adapt the control-system design. However, this is generally not a problem if the fan is properly designed in relation to the overall plant. During the initial design of the control system, dialogue with the process engineer or boiler designer will show whether or not surge protection will be required.

3.2.3 Final elements for draught control

Reference has already been made to the use of pitch-control in axial-flow fans to regulate the throughput of the machine. Other means of controlling flow are dampers, vanes or speed adjustment. Each of these devices has its own characteristics, advantages and disadvantages, and the selection of the controlling device which is to be used in a given application will be a trade-off between the technical features and the cost.

3.2.3.1 Types of damper

The simplest form of damper consists of a hinged plate that is pivoted at the centre so that it can be opened or closed across the duct. This provides a form of draught control but it is not very linear and it is most effective only near the closed position. Once such a damper is more than about 40–60% open it can provide very little additional control. Another form of damper comprises a set of linked blades across the duct (like a Venetian blind). Such multibladed dampers are naturally more expensive and more complex to maintain than single-bladed versions, but they offer better linearity of control over a wider range of operation.

The task of designing a control system for optimum performance over the widest dynamic range will be simplified if the relationship between the controller output signal and the resultant flow is linear. Although it is possible to provide the required characterisation within the control system, this will usually only be effective under automatic control. Under manual control a severely nonlinear characteristic can make it difficult for the operator to achieve precise adjustment.

It is possible to linearise the command–flow relationship under both manual and automatic control by the design of the mechanical linkage between the actuator and the damper. However, this requires careful design of the mechanical assemblies and these days it is generally considered simpler to build the required characterisation into the DCS. This approach provides a partial answer, but it should not be forgotten that such a solution is only effective under automatic control.

3.2.3.2 Vane control

The second form of control is by the adjustment of vanes at the fan inlet. These vanes are clearly visible near the centre of Figure 3.7 (which shows a centrifugal fan during manufacture). Such vanes are operated via a complex linkage which rotates all the vanes through the same angle in response to the command signal from the DCS.

3.2.3.3 Variable-speed drives

Finally, control of fan throughput can be achieved by the use of variable-speed motors (or drives). These may involve the use of electronic controllers which alter the speed of the driving motor in response to demand signals from the DCS or they can be hydraulic couplings or variable-speed gearboxes, either of which allows a fixed-speed motor to drive the fan at the desired speed. Variable speed drives offer significant advantages in that they allow the fan to operate at the optimum speed for the required throughput of air or gas, whereas dampers or vanes control the flow by restricting it, which means that the fan is attempting to deliver more flow than is required.

Figure 3.7 A centrifugal fan during manufacture
© Howden Sirocco Ltd. Reproduced with permission

3.3 Fuel systems

Fossil fuels that are burned in boilers can be used in solid, liquid or gaseous form, or a mixture of these. Naturally, the handling systems for these types of fuels differ widely. Moreover, the variety of fuels being burned is enormous. Solid fuels encompass a wide spectrum of coals as well as wood, the waste products of industrial processes, municipal and clinical waste and refuse-derived fuels. (The last are produced by shredding or grinding domestic, commercial and industrial waste material.) Liquid fuels can be heavy or light oil, or the products of industrial plant. Gas can be natural or manufactured, or the by-product of refineries.

Each of these fuels requires specialised handling and treatment, and the control and instrumentation has to be appropriate to the fuel and the plant that processes it.

3.3.1 Coal firing (pulverised fuel)

Although coal can be burned in solid form on grates, it is more usual to break it up before feeding it to the combustion chamber. The treatment depends on the nature of the coal. Some coals lend themselves to being ground down to a very fine powder (called pulverised fuel (PF)) which is then carried to the burners by a stream of air. Other coals are fed to impact mills which use flails or hammers to break up the material before it is propelled to the burners by an air stream. The type of mill to be used on a particular plant will be determined by the process engineers and it is the task of the control engineer to provide a system which is appropriate. To do this it is necessary to have some understanding of how the relevant type of mill operates.

Various types of pulverised-fuel mill will be encountered, but two are most commonly used: the pressurised vertical-spindle ball mill and the horizontal-tube mill.

3.3.1.1 Vertical-spindle ball mills

Figure 3.8 shows the operating principle of a typical ball mill, such as the Babcock 'E' mill. In this device, the coal that is discharged from the storage hoppers is fed down a central chute onto a table where it is crushed by rotating steel balls. Air is blown into the crushed coal and carries it, via adjustable classifier blades, to the PF pipes that transport it to the burners.

The air that carries the fine particles of coal to the burners is supplied from a fan called a 'primary-air fan'. This delivers air to the mill, which therefore operates under a pressure which is slightly positive with respect

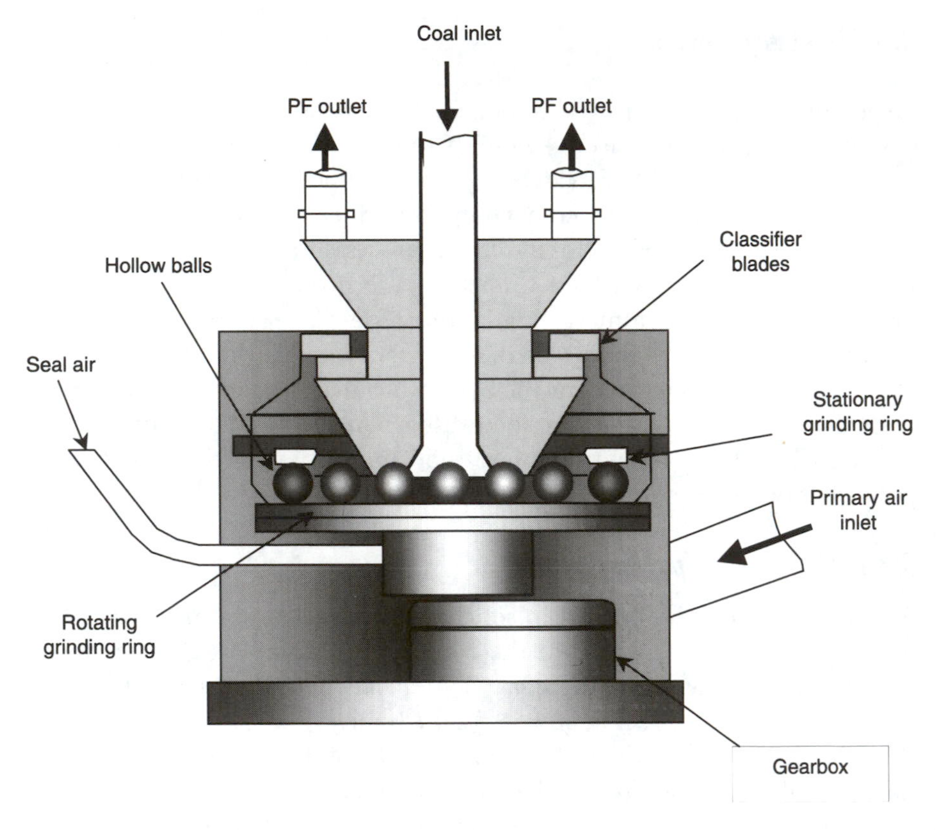

Figure 3.8 Pressurised ball mill

to the atmosphere outside. Because of this and because of its other constructional features, this type of mill is properly called a 'vertical-spindle, pressurised ball mill'. The air-supply system for this type of mill is discussed in more detail in Section 3.3.1.3.

3.3.1.2 Horizontal tube mills

In a tube mill (Figure 3.9) the coal is fed into a cage that rotates about a horizontal axis, at a speed of 18 to 35 rpm. This cage contains a charge of forged-steel or cast alloy balls (each of which is between 25 mm and 100 mm in diameter) which are carried up the sides of the cage by the rotation, until they eventually cascade down to the bottom, only to be carried up again. The coal is pulverised by a combination of the impact with these balls, attrition of adjacent particles and crushing between the balls and the cage and between one ball and another.

In this type of mill the crushed mixture is drawn out of the cage by a fan, which is called an exhauster. Because of this configuration, the tube

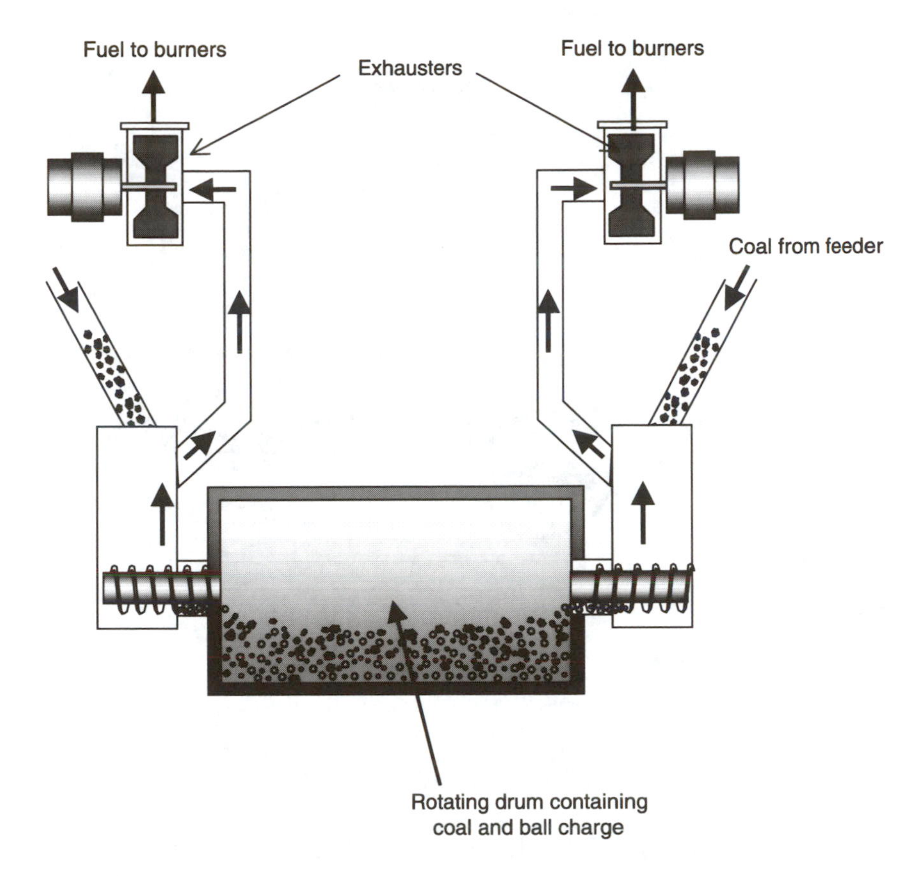

Figure 3.9 Horizontal tube mill

mill runs under a negative pressure, which prevents the fine coal dust from escaping (as it tends to do with a pressurised mill). However, the exhausters have to handle the dirty and abrasive mixture of coal and air that comes through the mill and they therefore require more frequent maintenance than the fans of a pressurised ball mill, whose function is merely to transport air from the atmosphere to the mill.

3.3.1.3 Air supply systems for mills

As stated above, the crushed coal in a pressurised ball mill is propelled to the burners by a stream of warm air. Figure 3.10 shows the arrangement for doing this: cool air and heated air are mixed to achieve the desired temperature. This temperature has to be high enough to partially dry the coal, but it must not be so high that the coal could overheat (with the risk of the coal/air mixture igniting inside the mill or even exploding while it is being crushed). The warm air is then fed to the mill (or a group of mills) by

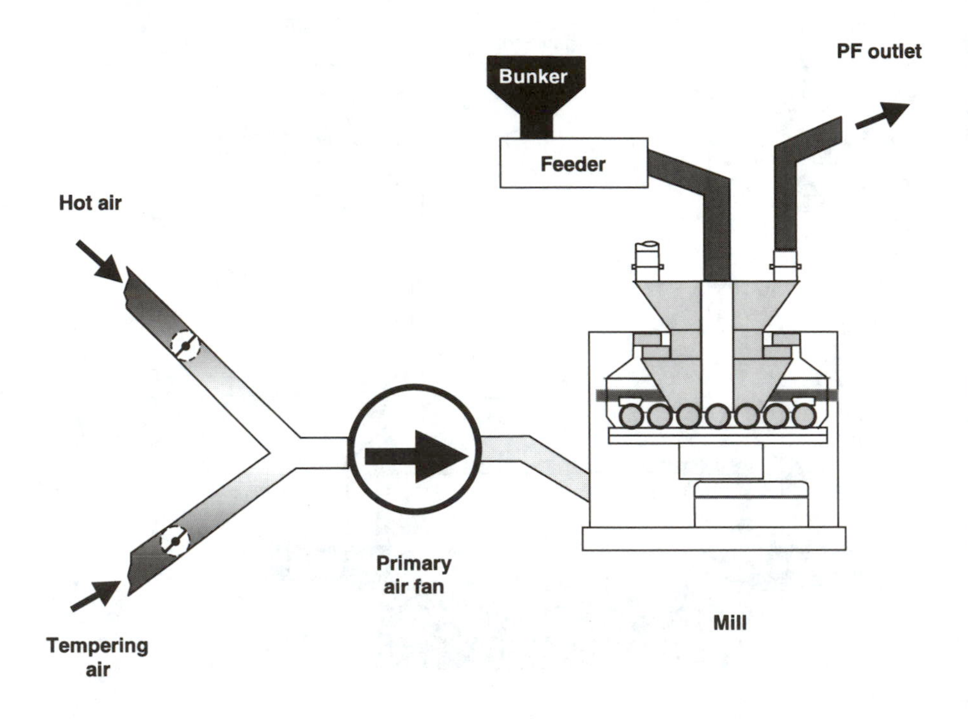

Figure 3.10 Primary air fan system for a ball mill

means of yet another fan, called a 'primary air fan'. It should be noted that the cooler of the two air streams is commonly referred to as 'tempering air' since, because it is obtained from the FD fan exhaust it may already be slightly warm, and its function is to temper the mixture.

Figure 3.11 shows the system that is used with a tube mill. Here, hot air and cold air are again mixed to obtain the correct temperature for the air stream but, because the mill in this case operates under suction conditions a primary air fan is not needed, and the cold air is obtained directly from the atmosphere. The warmed air mixture is again fed to the mill as 'primary air' but in addition a stream of hot air is fed to the feeder for transportation and drying purposes.

3.3.2 Oil-firing systems

In comparison with coal, oil involves the use of much less capital plant. On the face of it, it would appear that all that is required is to extract the oil from its storage tank and pump it to the burners. But in reality life is more complicated than that!

Proper ignition of oil depends on the fuel being broken into small droplets (atomised) and mixed with air. The atomisation may be achieved

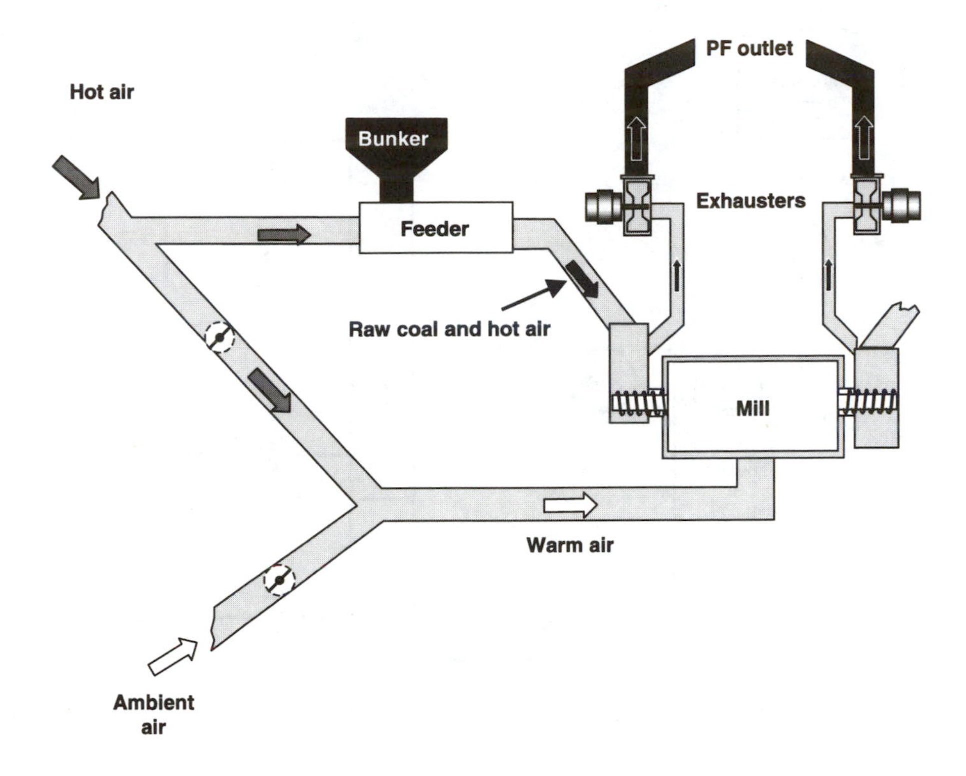

Figure 3.11 Suction mill air supply system

by expelling the oil through a small nozzle (a 'pressure jet'), or it may be achieved by the use of compressed air or steam.

The fuel oil itself may be light (such as diesel oil or gas oil), or it may be extremely viscous and tar-like (heavy fuel oil, commonly 'Bunker C'). The handling system must therefore be designed to be appropriate to the nature of the liquid. With the heavier grades of oil, prewarming is necessary, and to prevent it cooling and thickening the fuel is continually circulated to the burners via a recirculation system (shown schematically in Figure 3.12). The latter process is sometimes referred to as 'spill-back'. When a burner is not firing, the oil circulates through the pipework right up to the shut-off valve, which is mounted as close to the oil gun as possible.

From the point of the C&I engineer, the control systems involved with oil firing may include any or all of the following: controlling the temperature of the fuel, the pressure of the atomising medium, and the equalisation of the fuel pressures at various levels on the burner front.

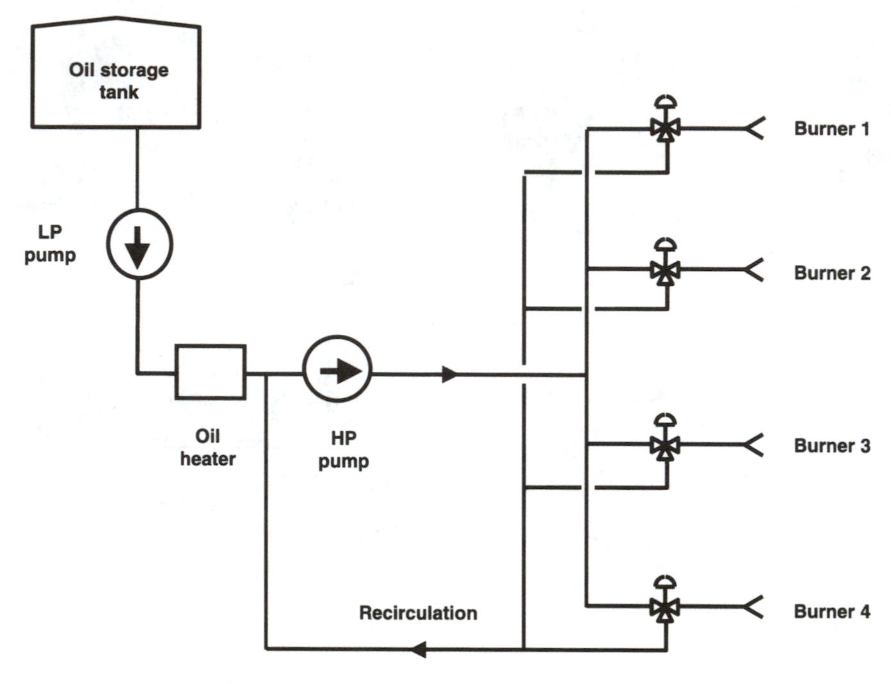

Figure 3.12 Simplified oil pumping, heating and recirculation system

3.3.3 Gas-firing systems

Although inherently simpler than either oil or coal-fired systems, gas-fired boilers have their own complexities. Any escape of gas, particularly into confined areas, presents considerable hazards, and great care must therefore be taken to guard against leakage, for example, from flanges and through valves. But natural gas is colourless, and any escape will therefore be invisible. Also, it is not safe to rely on odour to detect leakages. By the time an odour has been detected sufficient gas may have already escaped to present a hazard. It is therefore necessary for gas-leak detectors to be fitted along the inlet pipework wherever leakage could occur, and to connect these to a comprehensive, central alarm system.

It is also necessary to prevent gas from seeping into the combustion chamber through leaking valves. If gas does enter undetected into the furnace during a shut-down period, it could collect in sufficient quantities to be ignited either by an accidental spark or when a burner is ignited. The resulting explosion would almost certainly cause major damage and could endanger lives. (It should be noted that this risk is present with propane igniters such as those used with fuels other than oil.)

Protection against leakage into the furnace through the fuel-supply valves is achieved by the use of 'double-block-and-bleed' valve assemblies which provide a secure seal between the gas inlet and the furnace. The operation of this system (see Figure 3.13) is that before a burner is ignited both block valves are closed and the vent is open. In this condition any gas which may occupy the volume between the two block valves is vented to a safe place and it can therefore never develop enough pressure to leak past the second block valve. When start-up of the burner is required, a sequence of operations opens the block valves in such a way that gas is admitted to the burner and ignited safely.

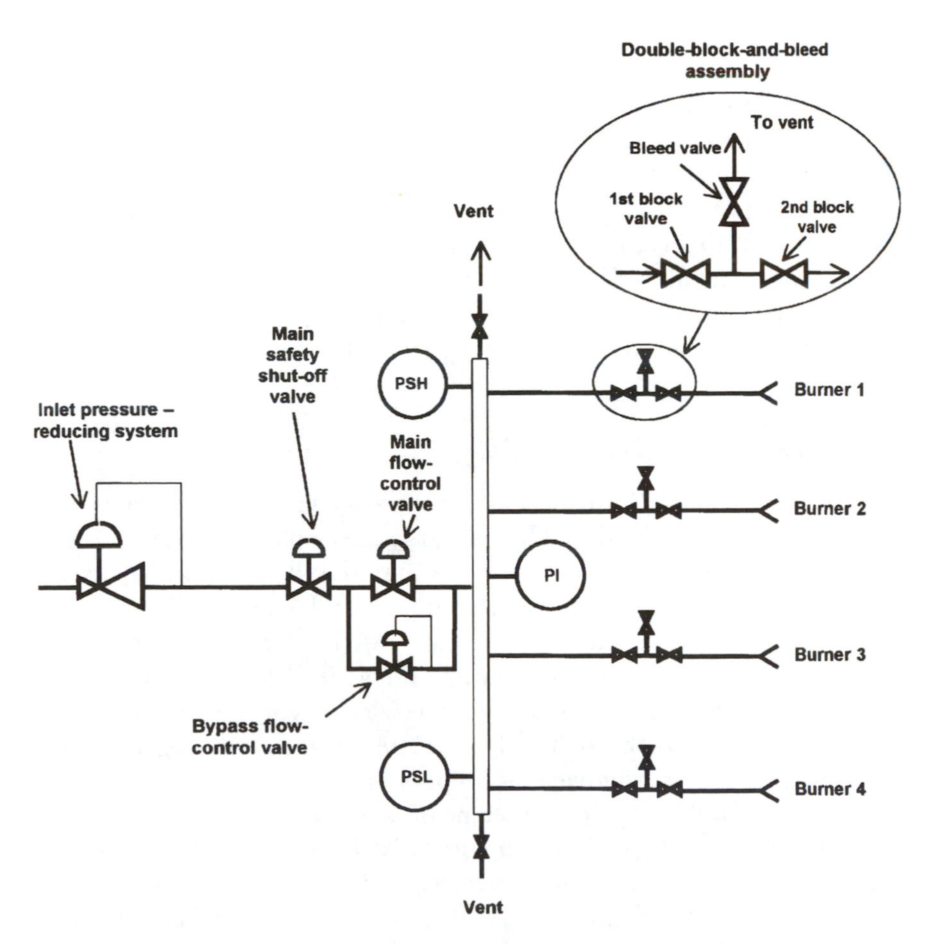

Figure 3.13 Simplified schematic of gas-firing system

3.3.4 Waste-to-energy plants

There has been a steady development of plants that incinerate waste material of various types and use the heat thus produced to generate electricity. Early units suffered from the unpredictable nature of the waste material and the severe corrosion resulting from the release of acidic compounds during the combustion process. But the problems have been largely overcome through the application of improved combustion systems and by better knowledge of the materials used in the construction of the plant.

Waste material may be obtained from any of several sources, including the following:

- municipal;
- industrial;
- clinical;
- agricultural.

The material may be burned after very basic treatment (shredding etc.) or it may be processed in some way, in which case the end result is termed refuse-derived fuel (RDF).

Several types of waste-to-energy plant are in existence, and we shall look at one of them, so that its nature and characteristics can be appreciated. Other plants will differ in their construction or technology, but from an operational point of view their fundamental characteristics will probably be quite similar to those described below.

3.3.4.1 The bubbling fluidised-bed boiler

Figure 3.14 shows the principles of a waste-to-energy plant based on the use of a bubbling fluidised-bed boiler. First, the incoming waste is sorted to remove oversized, bulky or dangerous material. The remainder is then carried by a system of conveyors to a hammer mill where it is broken down until only manageable fragments remain. After a separator has removed incombustible magnetic items, the waste is held in a storage building, from where it is removed as required by a screw conveyor and transferred via another conveyor to the boiler. Immediately before entering the boiler, nonferrous metals are removed by a separator.

The boiler itself comprises a volume of sand which is kept in a fluidised state by jets of air. A portion of dolomite is added to the sand to assist in the reduction of corrosion and to reduce any tendency of the sand and fuel to coalesce (a process known as 'slagging'). After the sand/dolomite mixture has been heated by a system of start-up burners, combustible

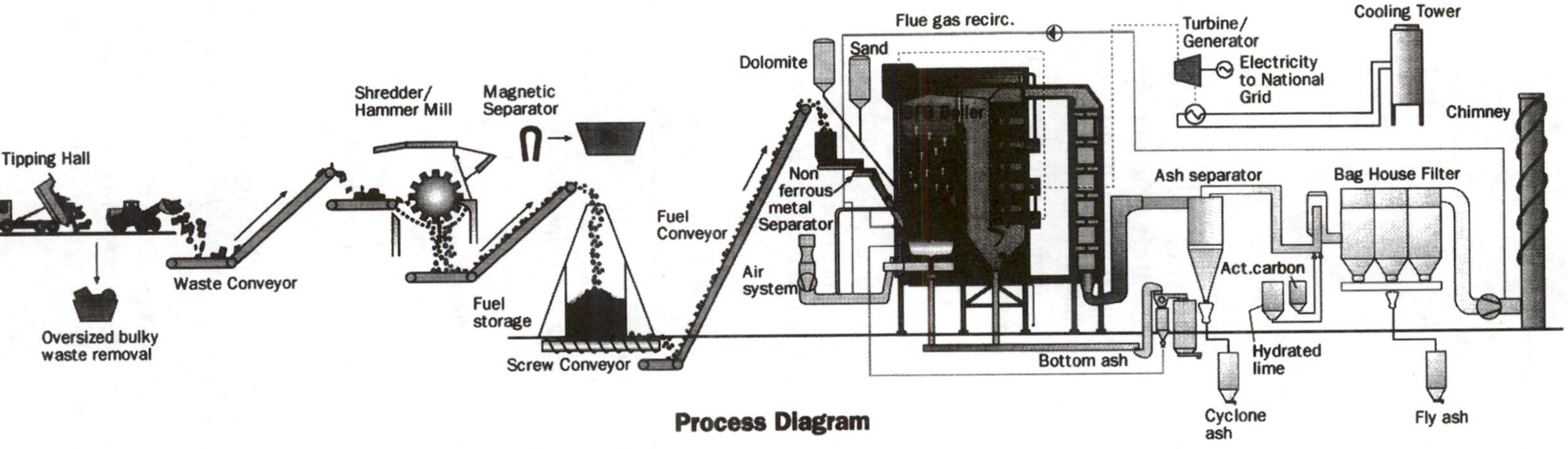

Figure 3.14 Waste-to-energy plant based on bubbling fluidised-bed boiler

waste material added to it ignites. The heat released is used to generate steam in a way that is similar to conventional boilers such as those described in Chapter 2.

3.4 Igniter systems

Whatever the main fuel of the boiler may be, it is necessary to provide some means of igniting it. A variety of igniters are used, but most modern systems comprise a means of generating a high-energy electric spark which lights a gas or light-oil supply which in turn lights the main fuel.

In addition to igniting the fuel, the igniter may sometimes be used to ensure that the fuel remains alight under conditions where it may otherwise be extinguished. This is referred to as providing 'support' for the main burner.

Like many aspects of power-station burner operations, the requirements for igniters are defined in standards such as those developed by the National Fire Protection Association (e.g. NFPA 8502:95 [1]). In these standards, igniters are divided into three categories each of which is defined in detail. In essence, the three classes have the following characteristics.

Class 1: An igniter providing sufficient energy to raise the temperature of the fuel and air mixture above the minimum ignition temperature, and to support combustion, under any burner light-off or operating conditions. Such igniters generally have a capacity of more than 10% of the full-load capacity of the main burner that it is igniting. This class is also referred to as a 'Continuous Igniter'.

Class 2: (also referred to as an 'intermittent igniter'): Capable of lighting the fuel *only under a defined range of light-off conditions.* Such igniters have a capacity generally between 4% and 10% of the full-load burner input and may also be used to support combustion of the fuel at low loads or under a defined range of adverse operating conditions.

Class 3: Small igniters, generally applied to gas or oil burners. These igniters are capable of lighting the fuel *only under a defined range of conditions and may not be used for support purposes.* Two types of Class 3 igniter are defined: interrupted igniters (not usually exceeding 4% of the main burner fuel input energy), whose operation is automatically stopped when a set time has expired after the first ignition; and direct electric igniters which have enough energy to ignite the main fuel.

The type of igniter in use will define the methods of operation of the burner and the sequences that are to be employed in the associated burner-management system.

3.5 Burner-management systems

Safe operation of the burner and its associated igniter must be ensured and in most cases this requires the use of a sophisticated burner-management system (BMS). In outline, these systems include a means of monitoring the presence of the flame and a reliable method and procedure for operating the associated fuel valves in a sequence that provides safe ignition at start-up and safe shut-down, either in the event of a fault or in response to an operator command.

The procedure for lighting a burner depends, first, on checking that it is safe to light it at all. This means that, if no other burner is firing, confirmation has been received that any flammable mixtures have been exhausted from the furnace by means of a purge. Such a purge involves the operation of FD and ID fans for a defined time, so that a certain volume of air has passed through the furnace. (In a coal-fired boiler the flow rate through the furnace must be at least 40% of the full-load volumetric air flow.)

Once confirmation has been received that the furnace purge is complete (or if other burners are already firing), ignition of the burner will depend on the successful operation of some form of igniter or pilot and, once the main burner has been successfully lit, its operation must be continuously monitored, because an extinguished flame may mean that unburned fuel is being injected into the combustion chamber. If such fuel is subsequently ignited it may explode.

Once a burner has ignited, the BMS must ensure that safe operation continues, and if any hazard arises the system must shut off the burner, and if necessary, trip the entire boiler.

On shut-down of a burner, steps must be taken to ensure that any unburned fuel is cleared from the pipework. This procedure is known as scavenging, and in an oil burner it may involve blowing compressed air or steam through the pipework and burner passages. Such procedures are defined in codes such as NFPA 8502-95.

Each component of the BMS is vital to the safety of the plant and to the reliability of its operation, but the most onerous responsibility rests with the flame detector: an electronic device which is required to operate in close proximity to high-energy spark ignition systems, and in conditions of extreme heat and dirt. Moreover, it must provide a reliable indication of the presence or absence of a particular flame in the presence of many others and it must discriminate between the energy of the flame and high levels of radiant energy from hot refractory materials and pipes. The sighting of the flame may also be affected by changes in flame pattern over a wide range of operating conditions, and it may also be obscured by swirling smoke, steam or dust.

Safe operation of the boiler depends on proper design of the BMS, including the flame scanner, and on careful siting of the scanner so that it provides reliable and unambiguous detection of the relevant flame under all operational conditions. After installation, the system can be expected to perform safely and reliably *only* if constant and meticulous attention is paid to maintenance. This important matter is all too often ignored, and the inevitable result is that the system malfunctions, leading to failure to ignite the fuel, which may in turn delay start-up of the boiler. In the extreme, malfunctions could even endanger the safety of the plant if they result in fuel being admitted to the combustion chamber without being properly ignited. A properly designed BMS will not allow this to happen, but if repeated malfunctions occur it is not unknown for operators to ignore the warning signs and even to override safety systems. In such cases it is usual to blame the BMS and/or the flame monitors, which could be fully functional if they were not misused or badly maintained. This important subject is discussed in greater depth in Chapter 5.

3.6 Gas turbines in combined-cycle applications

In the combined-cycle plant, the heat used for boiling the water and superheating the steam is obtained from the exhaust of a gas turbine, as described in Chapter 2. In such plant, unless supplementary firing is used, the combustion process occurs entirely in the gas turbine. Where supplementary firing is used the relevant control systems take on many of the characteristics of the oil- or gas-firing systems discussed earlier in the present chapter.

3.7 Summary

So far, we have looked at the operation of the boiler and studied in outline the boiler's steam, water and gas circuits, and all the major items of plant required for their operation. With this understanding we can now look at the control and instrumentation systems associated with the plant. This survey will be structured in much the same way as the preceding chapters, starting with an overview of an important fundamental: the method by which the demand for steam, heat or electrical power is obtained. Afterwards, we shall see how this demand is transmitted to all the relevant sections of the plant so that the requirements are properly and safely addressed.

3.8 References

1 NFPA 8502-95: Standard for the prevention of furnace explosions/implosions in multiple burner boilers. National Fire Protection Association, Batterymarch Park, Quincy, MA, USA, 1995

Chapter 4

Setting the demand for the steam generator

4.1 Nature of the demand

The steam generated by the boiler may be used to drive a turbine in a thermal power-plant, or it may be delivered to an industrial process or a district-heating scheme (or it may be provided for a mixture of these uses). Alternatively, the primary purpose of the plant may be to incinerate industrial, domestic or clinical waste, with steam being generated as a valuable by-product, to drive a turbo-generator or to meet a heating demand. In each case, the factor that primarily determines the operation of the plant is the amount of steam that is required. Everything else is subsidiary to this, although it may be closely linked to it.

The determinant that controls all the boiler's operations is called the 'master demand'. In thermal power-plant the steam is generated by burning fuel, and the master demand sets the burners firing at a rate that is commensurate with the steam production. This in turn requires the FD fans to deliver adequate air for the combustion of the fuel. The air input requires the products of combustion to be expelled from the combustion chamber by the ID fans, whose throughput must be related to the steam flow. At the same time, water must be fed into the boiler to match the production of steam.

As stated previously, a boiler is a complex, multivariable, interactive process. Each of the above parameters affects and is affected by all of the others.

The way in which the master demand operates is determined both by the general nature of the plant (is it a power station, an incinerator or a provider of process steam?), and also by the way in which the boiler is configured within the context of the overall plant (is there only one boiler meeting the demand, or are several combined?). The nature of the master demand system depends on the type of plant within which the boiler operates, and it is therefore necessary to examine it separately for each type of application. In the following sections we shall deal with the master demand as used in the following classes of plant:

- power stations;
- combined heat and power (CHP) plants;
- Waste-to-energy (WTE) plants.

We shall see that although all of these require the boiler to be operated to generate steam, each has its own requirements and constraints.

4.2 Setting the demand in power-station applications

A boiler producing steam for an operating turbo-generator has to ensure that the machine continually delivers the required electrical energy to the load. With a combined-cycle gas-turbine plant it is frequently the case that the power generated by the gas turbines is adjusted to meet the demand, with the steam turbine making use of all of the waste heat from the turbines.

With all types of power-generating plant, however, the requirement for generation will be set, directly or indirectly, by the grid-control centre (or the 'central dispatcher'), and the amount of power that is generated will be related to the local or national demand at that time.

In national networks, power stations are linked together to generate electrical power in concert with one another. Together they must meet a demand that is made up of the combined needs of all the users that are connected to the system (domestic, commercial, agricultural, industrial etc.). The overall demand will vary from minute to minute and day to day in a way that is systematic or random, dictated by economic, operational and environmental factors. This pattern of use relates to the entire network, and the fact that a large number of power generators and users are linked via the network has little bearing on the overall demand, although the extreme peaks and troughs may well be smoothed out. The interlinking does, however, have operational implications. For example, a sudden failure of one generating plant will instantly throw an extra demand on the others.

In a cold or temperate climate the demand will be based predominately on the need for light, heat and motive power. In warmer climates and developed areas it will also be determined by the use of air-conditioning and, possibly, desalination plant (for drinking-water production).

Figure 4.1 shows how the total electrical demand on the United Kingdom's Grid system varies from hour to hour through the day, and from a warm summer day to a cold winter day. Clearly, in addition to being affected by normal working patterns, the demand is determined by the level of daylight and the ambient temperature, both of which follow basic systematic patterns but which may also fluctuate in a very sudden and unpredictable manner. Similar profiles can be developed for each country and will be determined by climate as well as the country's industrial and commercial infrastructure.

These days, the demand for electricity in a developed nation is also affected quite dramatically by television broadcasts. During a major sporting event such as an international football match, sudden upsurges in demand will occur at half-time and full time, when viewers switch on their kettles. In the UK this can impose a sudden rise in demand of as much as 2 GW, which is the equivalent to the total output of a reasonably large

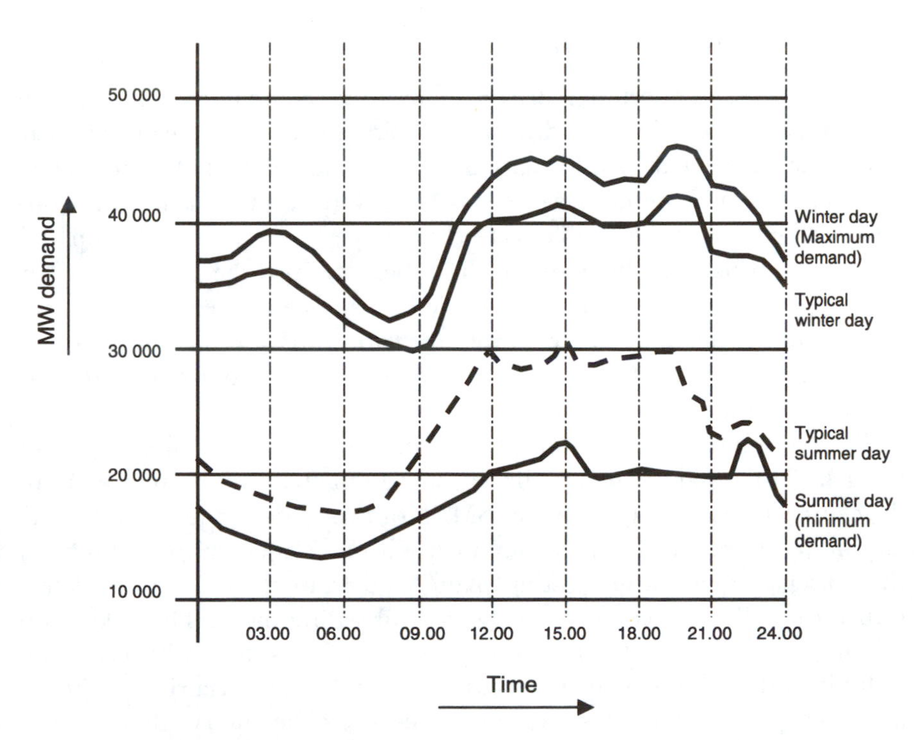

Figure 4.1 Typical electrical demand in the United Kingdom

power station. Such a pattern of usage can be predicted to within a few minutes, and audience predictions are routinely fed to the power-generation authorities on a daily basis to assist with the provision of adequate supplies. But if the result of the match requires 'extra time' playing there will be two further peaks before the pattern of consumption returns to normal an hour or so after the end of the match. This type of demand is obviously not predictable.

The Grid system has to be managed so that the demand for electricity is met within statutory limits at all times and under all conditions, and the available generating plant has to be used in the most economic manner.

Since the privatisation of the electricity supply industry in the UK, the generation of electricity is based on the demands of a trading system known as the Pool. This is briefly described below, because the operation of the Pool determines how each unit receives demand instructions. The subject is of critical importance because it governs the operation of the power plant and, ultimately, its demise. Although the following outline is based on the UK Pool, other countries use systems based on similar principles.

4.2.1 Operation of the UK Pool

At 10 a.m. each morning, British generating companies who wish to trade during the following day submit bids on behalf of each of their plants, accompanied by information on the capacity available from each plant and its operational parameters. These bids are then ranked nationally in a form of a league table, with the cheapest generator at the top and the most expensive at the bottom. This table is termed the 'merit order' for all the generating units that are capable of being connected to the system. The details of each day's merit order are transmitted to the body responsible for operating the Grid system, the National Grid Control Centre, and to the body responsible for the trading systems.

The National Grid Control Centre determines a notional schedule of the generating plant that is available and, on the basis of this information, develops a system marginal price (SMP) for every 30-minute period of the day in question. The SMP is combined with a component which reflects the scarcity values of generating plant, and from these factors is determined the selling price for power, the 'pool selling price'. This takes into account the operational costs of the system, known as the 'uplift costs'. The uplift includes factors such as the cost of maintaining a security margin of available power above the demand, the cost of ancillary plant that is required to maintain voltage and frequency, transmission constraints etc.

The National Grid Control Centre examines the security of the transmission network that links all the participating generating plant and consumers, and then plans its operations to ensure that the entire system operates securely and efficiently.

At the end of this process, the Centre issues instructions to all power stations connected to the network, setting the generation demand for their units on a minute-by-minute basis. It is these commands which determine the earnings of the plant, and within any given plant the boiler and turbine must respond to them in the most efficient and reliable manner possible. The load allocated to a unit will be based both on the cost of generation and on the ability of the plant to respond to demand changes. Under the right conditions, a unit whose operational costs are high but which responds quickly will be as likely to receive a load as one which generates very cheaply but is slow to respond to changes in demand.

4.3 The master demand in a power-station application

The response of a boiler/turbine unit in a power station is determined by the dynamic characteristics of the two major items of plant. These differ quite significantly from each other. The turbine, in very general terms, is capable of responding more quickly than the boiler to changes in demand. The response of the boiler is determined by the thermal inertia of its steam and water circuits and by the characteristics of the fuel system. For example, a coal-burning boiler, with its complex fuel-handling plant, will be much slower to respond to changes in demand than a gas-fired one.

Also, the turndown of the plant (the range of steam flows over which it will be capable of operating under automatic control) will depend on the type of fuel being burned, with gas-fired units being inherently capable of operating over a wider dynamic range than their coal-fired equivalents.

The common factor in all these systems, however, is the master demand which, in addition to setting the firing rate, regulates the quantity of combustion air to match the fuel input, and the quantity of feed water to match the steam production. In the present chapter we shall examine the master system. Chapters 3, 5, 6 and 7 look at how the commands from the master system are acted upon by the fuel, draught, feed-water and steam-temperature systems.

The design of the master system is determined by the role which the plant is expected to play, and here three options are available. The demand signal can be fed primarily to the turbine (boiler-following control); or to the boiler (turbine-following control); or it can be directed to both (co-ordinated unit control). Each of these results in a different performance of the unit, in a manner that will now be analysed.

4.3.1 Boiler-following operation

With boiler-following control, the power-demand signal modulates the turbine throttle-valves to meet the load, while the boiler systems are modulated to keep the steam pressure constant. The principles of this system are illustrated in simplified form in Figure 4.2.

In such a system, the plant operates with the turbine throttle-valves partly closed. The action of opening or closing these valves provides the desired response to demand changes. Sudden load increases are met by opening the valves to release some of the stored energy within the boiler. When the demand falls, closing the valves increases the stored energy in the boiler.

In such a system the turbine is the first to respond to the changes. The boiler control system reacts after these changes have been made, increasing or reducing the firing to restore the steam pressure to the set value.

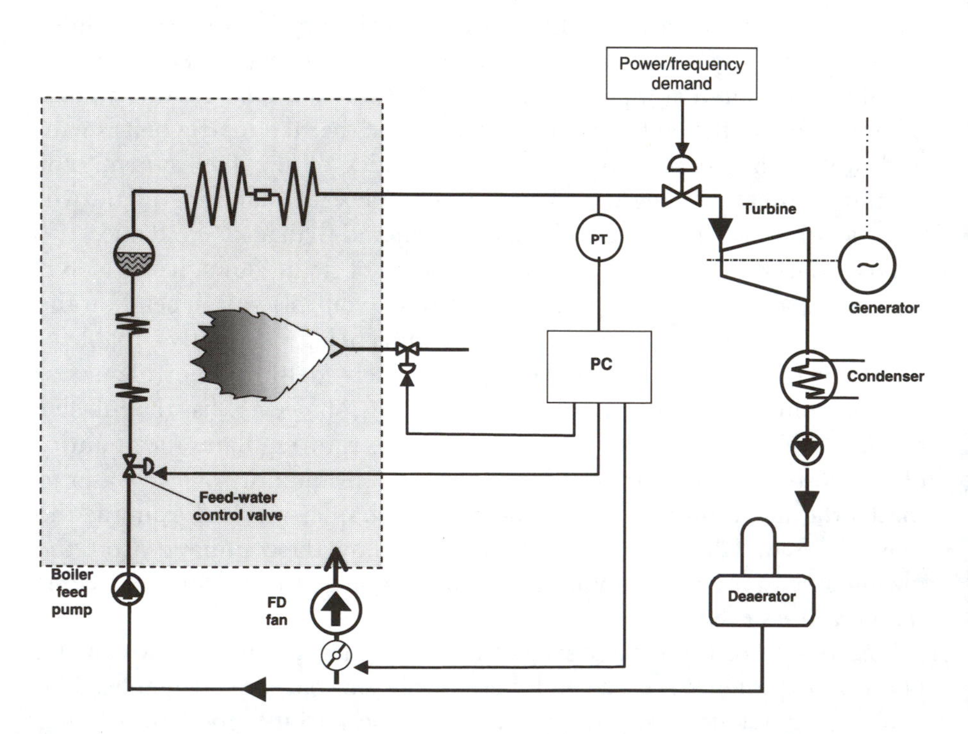

Figure 4.2 'Boiler following' system

4.3.2 Turbine-following operation

In the turbine-following system (Figure 4.3), the demand is fed directly to the boiler and the turbine throttle-valves are left to maintain a constant steam pressure. Particularly in the case of coal-fired plant, this method of operation offers slower response, because the turbine output is adjusted only after the boiler has reacted to the changed demand. However, the turbine-following system enables the unit to be operated in a more efficient manner and tuning for optimum performance is easier than with the boiler-following system. It is worth considering for large base-load power plant (where the unit runs at a fixed load, usually a high one, for most of the time), or with gas-fired plant where the response is comparatively rapid.

4.3.3 Co-ordinated unit control

With co-ordinated unit load control (Figure 4.4), the power demand is fed to the boiler and turbine in parallel, with various constraints built into each channel to recognise and allow for any dynamic constraints of the relevant plant. This is a sophisticated technique, which has come into its own with the development of powerful, fast, and versatile computer systems. It combines the best features of both the boiler-following and the

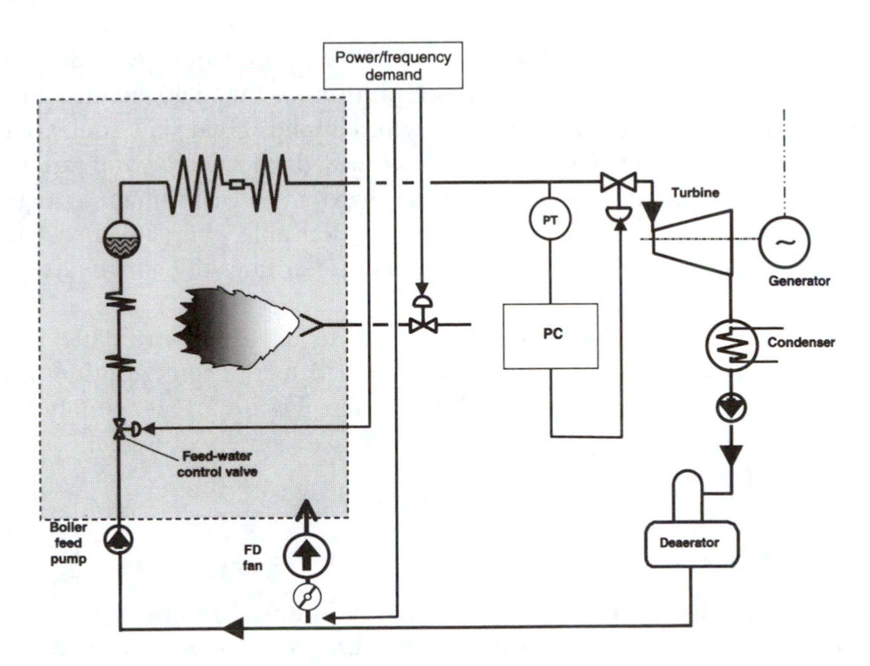

Figure 4.3 'Turbine following' system

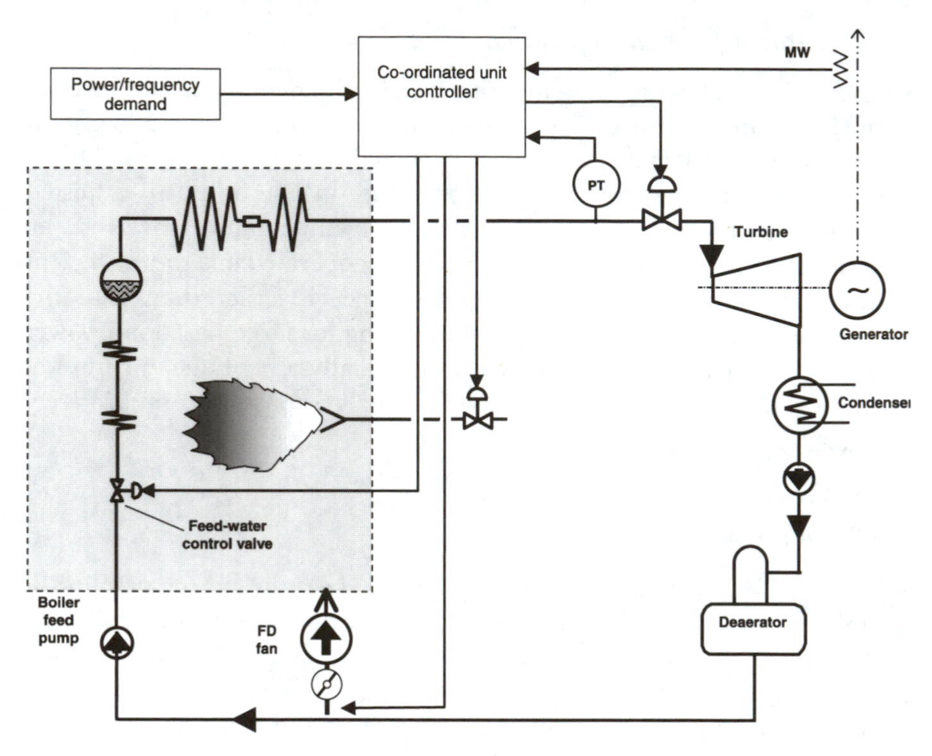

Figure 4.4 'Co-ordinated unit control' system

turbine-following systems. However, its design demands considerable knowledge of the characteristics and limitations of the major plant items. Also, commissioning of this type of system demands great skill and care if the full extent of the benefits is to be obtained. In particular, the rate-of-change of the demand signals, as well as the extent of their dynamic range, will need to be constrained to prevent undesirable effects such as the stressing of pipework because of excessively steep rates-of-change of temperature.

Because of the nature of its operation, the details of a co-ordinated unit load control system have to be finely matched to the configuration and characteristics of the plant to which it is fitted. Figure 4.4 shows only the overall principle of the system: in practice, a wide variety of system configurations are in use.

4.3.4 Relative performance

Of the three options described above, the co-ordinated unit load control system provides the best possible response to changes in demand. However, because it is so sophisticated its performance is heavily dependent on the accuracy of the many pieces of information on which its operation is based (such as thermal rate-of-change limits in the turbine). Unless it is regularly readjusted, it can suffer from an inability to recognise and deal with the steady deterioration in performance that inevitably occurs in each item of plant as it ages.

4.3.5 Comparing the response rates of the systems

As stated above, the co-ordinated unit load controller, when properly designed, commissioned and maintained, will provide the best possible response of the unit within the constraints of the plant itself.

Unfortunately, for many practical reasons it is not universally used. In older plant this type of master configuration may not be available or practical. In other cases economic or time constraints may have necessitated the use of a simpler régime (the co-ordinated unit system is expensive to design, and its commissioning requires much time and effort if the maximum benefit is to be gained from its capabilities).

Where the co-ordinated unit system is not available, the choice lies between using boiler-following or turbine-following control. Although they both orchestrate the operation of the boiler and the turbine to meet changes in demand, the performances of these configurations differ very considerably from each other, as is explained below.

4.3.5.1 Response of the boiler-following system

When a change occurs in its steam-flow demand, a boiler has to overcome its own thermal inertia before it can match the change. Therefore, by using the turbine's ability to respond more rapidly, the boiler-following system provides a better response to load changes than the turbine-following system. After the turbine has responded to the change in demand, the boiler is commanded to follow on, correcting the steam-pressure error as quickly as it can.

However, such rapid response is only available for small-scale demand changes, that is, changes that are within the capacity of the allowable range of pressure-drop across the throttle valve. Also, the rapid response is obtained at a cost. When operated in this way, the efficiency of the unit is inevitably reduced because of the pressure that is dropped across the throttle valves. The losses are reduced by decreasing this pressure drop, but this also reduces the scope for meeting sudden changes in demand.

Another problem is that it is not easy to tune the control parameters of a boiler-following system to obtain optimum overall performance, mainly because of the interaction between steam pressure and steam flow that occurs as the turbine and boiler respond to changes.

Consider what happens when a sudden rise in demand occurs. The first response is for the throttle valves to be opened. This increases the power generated by the machine, but it also results in the boiler pressure falling, and when this happens the boiler control system reacts by increasing the firing rate. This is all right as far as it goes since, quite correctly, it increases the boiler steaming rate to meet the increase in demand. However, as the firing change comes into effect and the steam pressure rises, the amount of power that is being generated also increases. But as it has already been increased to meet the demand—and in fact may have already done so—the power generated can overshoot the target, causing the throttle valves to start closing again, which raises the boiler pressure . . . , and so on.

Various methods have been proposed to anticipate these effects, but these tend to increase the complexity of the system, and therefore its cost, with questionable long-term benefit.

4.3.5.2 Response of the turbine-following system

In the simplest version of the turbine-following system the boiler firing rate, and the rate of air and feed-water admission etc., are all fixed (or, at least, held at a set value, which may be adjusted from time to time by the boiler operator), and the turbine throttle valves are modulated to keep the steam pressure constant. However, when the fuel, air and water flows of a boiler are held at a constant value the amount of steam that is generated will not, in general, remain constant, mainly because of the inevitable variations that will occur in parameters such as the calorific value of the fuel, the temperature of the feed water etc. In the simple turbine-following system, these variations are corrected by modulation of the turbine throttle valve to maintain a constant steam pressure, but this results in variations in the power generated by the turbine.

Because the steam-generation rate of its boiler is not automatically adjusted to meet an external demand, a plant operating under the control of a simple turbine-following system will generate amounts of power that do not relate to the short-term needs of the grid system. Such a plant is therefore incapable of operating in a frequency-support mode, although this mode of operation may be used where it is not easy, or desirable, to adjust the fuel input, for instance in industrial waste-incineration plants.

Figure 4.3 shows the preferred option for other types of plant, with the boiler firing rate (and the input of air and water) being set by the demand on the unit. Changes in this demand therefore change the boiler's firing rate, and a controller then modulates the turbine throttle valve to keep the steam pressure constant. This technique closes the loop around the power, but it does so by directing the changes in demand to the boiler first, relying on the resulting changes in pressure to change the amount of power that is generated. As might be expected, because of the slow reaction time of the boiler, this results in a slower response to load changes than that of the boiler-following system.

4.4 Load demand in combined heat and power plants

Reference has already been made to the use of gas turbines in combined-cycle installations. This is a particular example of a 'co-generation' scheme: a term applied to dual-purpose plants where heat which would otherwise be wasted from one process is used in another. In the case of CCGT plant, the heat exhausted from a gas turbine is used to generate steam. In CHP plant, heat from a power station is used in another process. The heat may be taken from the power plant as steam extracted from the turbine, or it may be the heat abstracted from the condensate.

Co-generation plants are either 'topping' or 'bottoming' systems. With the former, the first priority is to generate electricity, and as much use as possible is made of the heat that would otherwise be wasted in the process. With the latter, waste heat from some industrial process is used to generate electricity via a steam generator and turbine.

A steam generator employed in a CHP plant has to serve two masters: the need for heat, and the demand for electricity generation. In most cases the former predominates, because the entire *raison d'être* of the plant was probably the need to serve a community or an industrial plant, and the plant's ability to generate electricity is of secondary importance (even though, as a spin-off, it is extremely valuable).

For this reason, the development of a truly effective master-demand signal for a CHP plant is much more complex than it is with a plant whose only function is to generate electricity. The needs of all the users have to be taken into consideration, as must the cost of the steam, heat and electricity that is produced. Furthermore, it is possible that the way in which the master demand is configured may need to be modified at some time over the life of the plant because of changes in fuel prices or alterations in the requirements of the industrial, commercial or domestic complexes which benefit from the process.

The wide range of possibilities of interconnecting the various systems in CHP plant gives rise to very diverse methods of organising the master demand. Configuring a master-demand signal that takes all the requirements into account ought not to be a significant problem, bearing in mind the power and flexibility that is offered by the modern DCS, but the difficulty is to obtain enough data on these requirements, and then to ensure that the information is correct. Quite often it seems that, even if the options might have been considered at some time, reconciling the various requirements has proved to be intractable and so a cheap and simple compromise has been employed. This may be reasonably effective, and the plant that is so developed continues to generate heat and power for days on end, the response to changes in demand seems adequate, and the operational staff are unprepared to alter anything for fear of rocking the boat.

None of this alters the fact that the expenditure of a few more days (or even weeks) of effort in front-end definition could have yielded, over the operational life of the plant, efficiency and performance improvements that would have amply recovered the cost of development.

4.5 Waste-to-energy plants

The design of master-demand signals of waste-to-energy plants, as described in Chapter 3, requires very careful attention. The requirements in this area are somewhat similar to those relating to CHP plants, the reconciliation of differing operational requirements.

The design of the system may favour consumption of the waste material, with electricity generation treated as a useful and revenue-earning by-product, or it may try to maximise the power-generation capability of the plant. In both cases, however, it is important to recognise the special characteristics of the plant, in particular, the boiler response.

Because of the nature of its complex fuel-handling system (see Figure 3.14), a waste-to-energy plant cannot be expected to be very responsive to demand changes. Therefore it is largely impractical to consider the application of advanced control logic to the master-demand system for this type of plant (although various attempts have been made to do so). The most cost-effective solution is to apply a simple boiler-following system as described in Section 4.3.1. The reduction in efficiency is negligible (and even somewhat academic, since not only is the fuel in this type of plant easily obtained, but also the user is paid for consuming it!). Also, the difficulties of tuning the system (due to the interaction between the steam generator and the steam user) are less of a problem in this type of installation, because of the very different dynamic responses of the turbine and boiler. The difference between the slow response of the boiler and the quick response of the turbine also simplifies decoupling one from the other in the optimisation process.

These factors ease the selection of a master system in waste-to-energy plants. A basic boiler-following design provides speed and simplicity of commissioning, and usually performs adequately.

4.6 Summary

In this chapter we have seen how a 'master demand' signal is generated in respect to the nature of the duties that the plant is designed to undertake. This signal is responsible for ensuring that the boiler reacts to changes in demand, and it must also co-ordinate the operation of each of the subsidiary systems. The main areas involved in this process are the combustion and draught systems, the feed-water system and the steam-temperature control system.

In the next chapter we shall see how the combustion and draught systems of a boiler react to the demands of the master signal to produce the required firing rate and how the supply of air keeps in step with the changes to produce the correct conditions for the combustion of the fuel. In addition, we shall see how the inlet and exhaust fans are regulated to maintain the correct pressure in the furnace while all this is going on.

Chapter 5

Combustion and draught control

When considering fired boilers and heat-recovery steam generators it is clear that in the areas of their steam and water circuits there are many similarities between them (although the HRSG may have two or more pressure systems). But when the systems for controlling the heat input are examined, the two types of plant take on altogether different characteristics. The reason for this is fundamental: within the HRSG, no actual combustion process is involved, since all the heat input is derived from the gas-turbine exhaust (except where supplementary firing is introduced between the gas turbine and the HRSG). The subject of combustion control, which we shall be examining in this chapter, is therefore only relevant to fired plant.

Naturally, in a fired boiler the control of combustion is extremely critical. In order to maximise operational efficiency combustion must be *accurate*, so that the fuel is consumed at a rate that exactly matches the demand for steam, and it must be executed *safely*, so that the energy is released without risk to plant, personnel or environment. (The amount of energy involved in a power plant is considerable: in each second of its operation a large boiler releases around a billion joules, and in a process of this scale the results of an error can be catastrophic.)

In this chapter we shall see how the combustion process is controlled to meet the two objectives defined in the previous paragraph. We shall also examine the subsidiary systems that maintain the correct operational conditions in the fuel-handling plant of coal-fired boilers.

5.1 The principles of combustion control

In Chapter 3 we saw that the theoretically perfect combustion of a fuel requires the provision of exactly the right amount of air needed for complete combustion of the fuel. For the boiler as a whole this means that the total amount of air being delivered to the combustion chamber at any instant matches the total amount of fuel entering that chamber at that time. For an individual burner it means that the fuel and air being delivered to the burner are always in step with one another.

On the surface, therefore, it would appear that the matter of combustion control merely involves keeping the fuel and air inputs in step with each other, according to the demands of the master, and if this were true this role would be adequately addressed by a straightforward flow ratio controller. Unfortunately, when the realities of practical plant are involved, the situation once again becomes far more complex than this simple analysis would suggest.

When the relationship between the fuel and air flowing at any instant into the furnace is chemically ideal for combustion, the relationship between the two flows is known as the stoichiometric fuel/air ratio. However, as stated earlier, it is usually necessary to operate at a fuel/air ratio that is different from this theoretically optimal value, generally with a certain amount of excess air. All the same, even though more than the theoretical amount of air has to be provided, any overprovision of air reduces the efficiency of the boiler and results in undesirable stack emissions, and must therefore be limited.

The reduction in efficiency is due to losses which are composed of the heat wasted in the exhaust gases and the heat which is theoretically available in the fuel, but which is not burned. As the excess-air level increases, the heat lost in the exhaust gases increases, while the losses in unburned fuel reduce (the shortage of oxygen at the lower levels increasing the degree of incomplete combustion that occurs). The sum of these two losses, plus the heat lost by radiation from hot surfaces in the boiler and its pipework, is identified as the total loss.

Figure 5.1 shows that operation of the plant at the point identified at 'A' will correspond with minimum losses, and from this it may be assumed that this is the point to which the operation of the combustion-control system should be targeted. However, in practice air is not evenly distributed within the furnace. For example, operational considerations require that a supply of cooling air is provided for idle burners and flame monitors, to prevent them being damaged by heat from nearby active burners and by general radiation from the furnace. Air also enters the

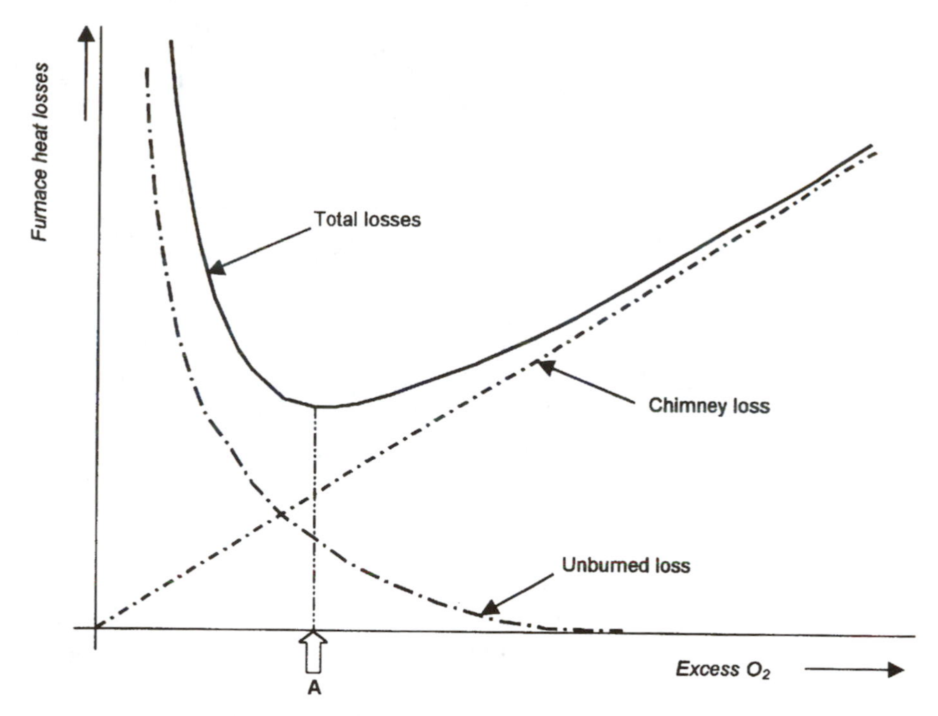

Figure 5.1 Heat losses in a furnace

combustion chamber through leaks, observation ports, soot-blower entry points and so on. The sum of all this is referred to as 'tramp air' or 'setting leakage'. If this is included in the total being supplied to the furnace, and if that total is apportioned to the total amount of fuel being fired, the implication is that some burners (at least) will be deprived of the air they need for the combustion of their fuel. In other words, the correct amount of air is being provided in total, but it is going to places where it is not available for the combustion process.

Operation of the firing system must take these factors into account, and from then on the system can apportion the fuel and air flows. If these are maintained in a fixed relationship with each other over the full range of flows, the amount of excess air will be fixed over the entire range.

5.1.1 *A simple system: 'parallel control'*

The easiest way of maintaining a relationship between fuel flow and air flow is to use a single actuator to position a fuel-control valve and an air-control damper in parallel with each other as shown in Figure 5.2. Here, the opening of an air-control damper is mechanically linked to the opening of a fuel control valve to maintain a defined relationship between

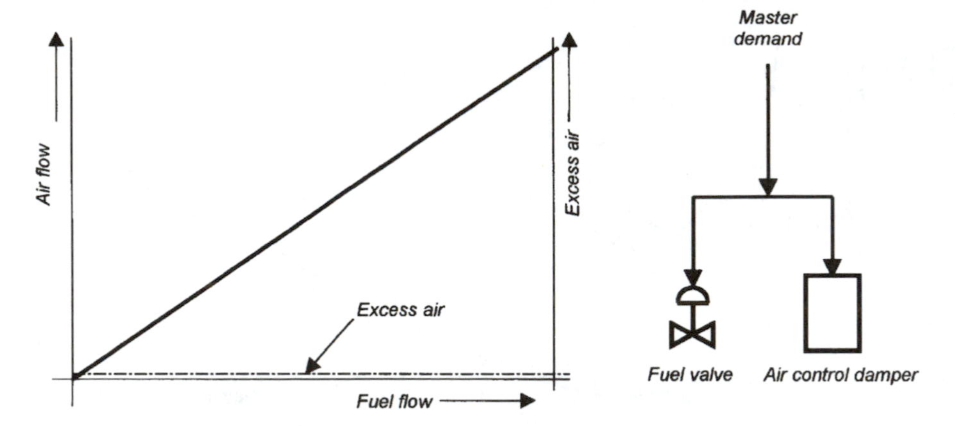

Figure 5.2 Simple 'parallel' control

fuel flow and air flow. This system is employed in very small boilers, and a variant allows a non-linear relationship between valve opening and damper opening to be determined by the shape of a cam, with a range of cams offering a variety of relationships.

Although this simple system may be quite adequate for very small boilers burning fuels such as oil or natural gas, its deficiencies become increasingly apparent as the size of the plant increases.

One limitation of the system is that it assumes that the amount of fuel flowing through the valve and the quantity of air flowing past the damper will remain constant for a given opening of the respective devices. In practice, if a valve or damper is held at a given opening, the flow past it will change as the applied pressure changes. Furthermore, the flow will also be affected by changes in the characteristics of the fuel and air, notably their densities.

Another problem is that the response times of the fuel and air systems are never identical. Therefore, if a sudden load-change occurs and the two controlling devices are moved to predetermined openings, the flows through them will react at different rates. With an oil-fired boiler, a sudden increase in demand will cause the fuel flow to increase quickly, but the air system will be slower to react. As a result, if the fuel/air ratio was correct before the change occurred, the firing conditions after the change will tend to become fuel-rich until the air system has had time to catch up. This causes characteristic puffs of black smoke to be emitted as unburned fuel is ejected to the chimney.

On a load decrease the reverse happens, and the mixture in the combustion chamber becomes air-rich. The resulting high oxygen content could lead to corrosion damage to the metalwork of the boiler, and to unacceptable flue-gas emissions.

5.1.2 Flow ratio control

The first approach to overcoming the limitations of a simple 'parallel' system is to measure the flow of the fuel and the air, and to use closed-loop controllers to keep them in track with each other, as shown by the two configurations of Figure 5.3.

In each of these systems the master demand (not shown) is used to set the quantity of one parameter being admitted to the furnace, while a controller maintains an adjustable relationship between the two flows (fuel and air). Either of the flows can be selected to be the one that responds directly to the master and, in Section 5.1.2.1, we shall see the different effects that result when fuel flow or air flow is used in this way.

In the system shown in Figure 5.3*a* a gain block or amplifier in one of the flow-signal lines is used to adjust the ratio between the two flows. As the gain (g) of this block is changed, it alters the slope of the fuel-flow/air-flow characteristic, changing the amount of excess air that is present at each flow. Note that when the gain is fixed, the amount of excess air is the same for all flows, as shown by the horizontal line.

In practice, this situation would be impossible to achieve, since some air inevitably leaks into the furnace, with the result that the amount of excess air is proportionally greater at low flows than high flows. This causes the excess-air line to curve hyperbolically upwards at low flows (much as is shown in Figure 5.3*b*). Practical burner requirements demand that the quantity of air should always be slightly greater than that which the theoretical stoichiometric ratio would dictate. The characteristic would therefore not pass through the origin of the graph as is shown in Figure 5.3*a*.

Figure 5.3*b* shows a different control arrangement working with the same idealised plant (i.e. one with no air leaking into the combustion chamber). Here, instead of a gain function, a bias is added to one of the signals. The effect of this is that a fixed surfeit of air is always present and this is proportionally larger at the smaller flows, with the result that the amount of excess air is largest at small flows, as shown. Changing the bias signal (b) moves the curve bodily as shown.

Each of these control configurations has been used in practical plant, although the version with bias (Figure 5.3*b*) exacerbates the effects of tramp air and therefore tends to be confined to smaller boilers. The arrangement shown in Figure 5.3*a* therefore forms the basis of most practical fuel/air ratio control systems.

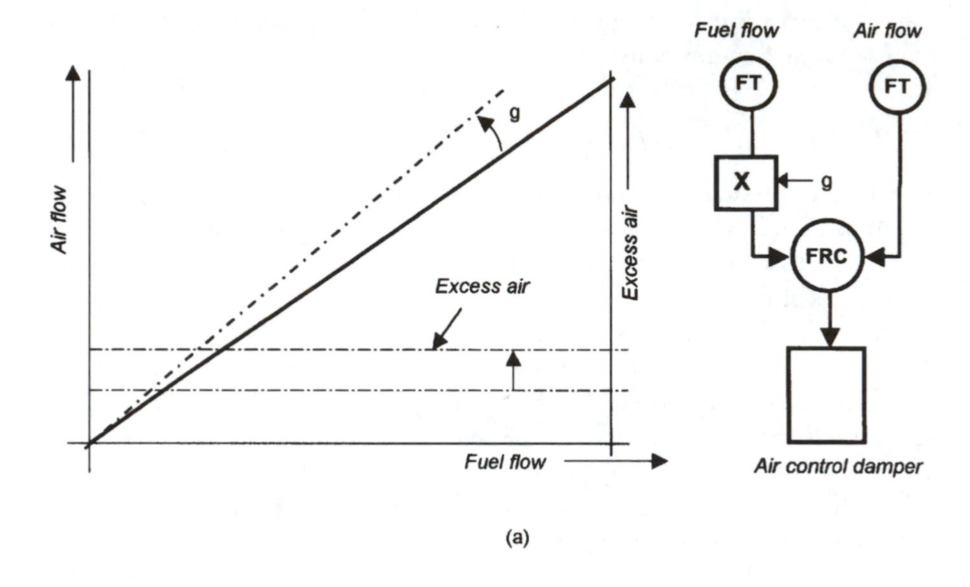

(a)

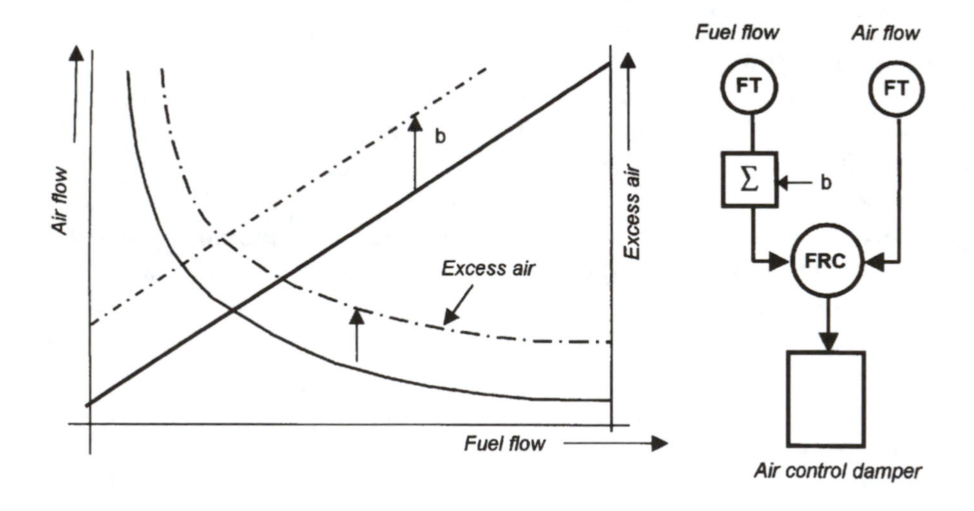

(b)

Figure 5.3 Fuel/air ratio control
a Gain adjustment of fuel/air ratio
b Bias adjustment of fuel/air ratio

In these illustrations it has been assumed that the master demand is fed to the fuel valve, leaving the air-flow controller to maintain the fuel/air ratio at the correct desired value. When this is done, the configuration is known as a 'fuel lead' system since, when the load demand changes, the fuel flow is adjusted first and the controller then adjusts the air flow to match the fuel flow, after the latter has changed.

It doesn't have to be done this way. Instead, the master demand can be relayed to the air-flow controller, which means that the task of maintaining the fuel/air ratio is then assigned to the fuel controller. For obvious reasons this is known as an 'air-lead' system.

5.1.2.1 Comparing the 'fuel-lead' and 'air-lead' approaches

Of the two alternatives described above, the fuel-lead version will provide better response to load changes, since its action does not depend on the slower-responding plant that supplies combustion air to the furnace. However, because of this, the system suffers from a tendency to produce fuel-rich conditions on load increases and fuel-lean conditions on decreases in the load. Operating in the fuel-rich region raises the risk of unburned fuel being ignited in an uncontrolled manner, possibly causing a furnace explosion. Whereas operating with too much excess air, while not raising the risk of an uncontrolled fire or an explosion, does cause a variety of other problems, including back-end corrosion of the boiler structure, and undesirable stack emissions.

The air-lead system is slow to respond because it requires the draught plant to react before the fuel is increased. Although this avoids the risk of creating fuel-rich conditions as the load increases, it remains prone to such a risk as the load decreases. However, the hazard is less than for the fuel-lead system.

A further limitation of these systems (in either the fuel-lead or air-lead version) is that they offer no protection against equipment failures, since these cannot be detected and corrected without special precautions being taken. For example, in the fuel-lead version, if the fuel-flow transmitter fails in such a way that it signals a lower flow than the amount that is actually being delivered to the furnace, the fuel/air ratio controller will attempt to reduce the supply of combustion air to match the erroneous measurement. This will cause the combustion conditions to become fuel-rich, with the attendant risk of an explosion. Conversely, if the fuel-flow transmitter in the air-lead system fails low, the fuel controller will attempt to compensate for the apparent loss of fuel by injecting more fuel into the furnace, with similar risks.

These are just some of the failure characteristics which the basic system design cannot address. Although the self-diagnostic features incorporated in modern transmitters can be arranged to raise an alarm and trip the burners, or operate the plant in a protected mode, until the fault has been corrected, it would be preferable to employ a system which has greater inherent abilities to deal with failures both in the plant and in its control and instrumentation equipment.

The so-called 'cross-limited' combustion control system addresses these factors in a very comprehensive way, as described in the following section.

5.1.3 Cross-limited control

Figure 5.4 shows the principles of the cross-limited combustion control system. Individual flow-ratio controllers (7, 8) are provided for the fuel and air systems, respectively. Ignoring for the moment the selector units (5, 6) and the fuel/air ratio adjustment block (4), it will be seen that the master demand signal is fed to each of these controllers as the desired-value signal, so that the delivery of fuel and air to the furnace continually matches the load. Because fuel flow and air flow are each measured as part of a closed loop, the system compensates for any changes in either of these flows that may be caused by external factors. For this reason it is sometimes referred to as a 'fully metered' system. The effect of the fuel/air ratio adjustment block (4) is to modify the air-flow signal in accordance with the required fuel/air relationship.

So far, the configuration performs similarly to the basic systems shown in Figure 5.3. The difference becomes apparent when the maximum and minimum selectors are brought into the picture. Remembering the problems of the differing response-rates of the fuel and air supply systems, consider what happens when the master demand signal suddenly requests an increase in firing. Assume that, prior to that instant, the fuel and air controllers have been keeping their respective controlled variable in step with the demand, so that the fuel-flow and modified air-flow signals are each equal to the demand signal. When the master demand signal suddenly increases, it now becomes larger than the fuel-flow signal and it is therefore ignored by the minimum-selector block which instead latches onto the modified air-flow signal (from item 4). The fuel controller now assumes the role of fuel/air ratio controller, maintaining the boiler's fuel input at a value that is consistent with the air being delivered to the furnace. The air flow is meanwhile being increased to meet the new demand, since the maximum-selector block (6) has latched onto the rising master signal.

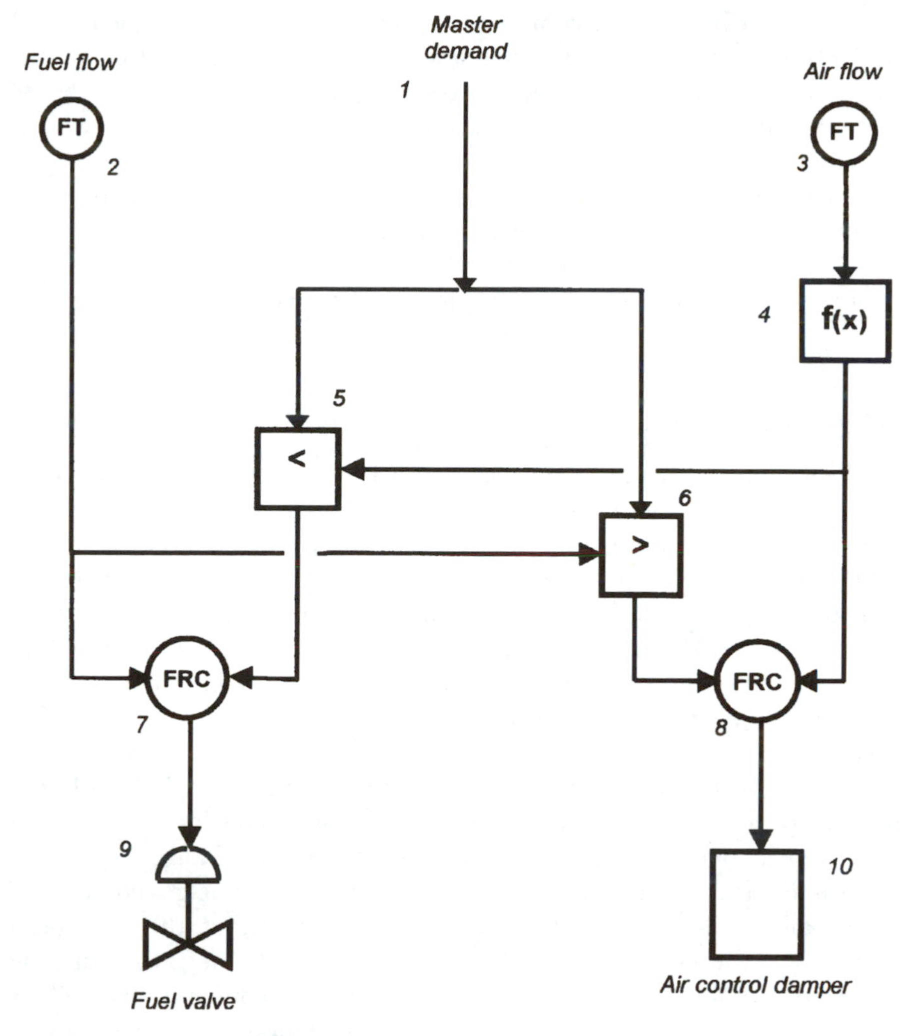

Figure 5.4 Basic cross-limited control system

On a decrease in load, the system operates in the reverse manner. The minimum-selector block locks onto the collapsing master and quickly reduces the fuel flow, while the maximum-selector block chooses the fuel-flow signal as the demand for the air-flow controller (8), which therefore starts to operate as the fuel/air ratio controller, keeping the air flow in step with the fuel flow.

Analysis of the system will show that it is much better able to deal with plant or C&I equipment failures. For example, if the fuel valve fails open, the air controller will maintain adequate combustion air to meet the quantity of fuel being supplied to the combustion chamber. This may result in overfiring but it cannot cause fuel-rich conditions to be created in the furnace. Similarly, if the fuel-flow transmitter fails low, although the fuel controller will still attempt to compensate for the apparent loss of fuel, the air flow controller will ensure that adequate combustion air is supplied.

The system cannot compensate for all possible failures, but it provides a much higher level of protection than any of the simpler systems described earlier, and when coupled with self-checking diagnostics and proper fault-detection techniques it provides a high degree of safety.

5.1.3.1 Using gas analysis to vary the fuel/air ratio

In the systems shown in Figures 5.3 and 5.4, the relationship between the fuel and air quantities is manually adjusted, either the gain or the bias is altered to change the combustion conditions. With such systems, if the adjustment factor is set wrongly, or if changes outside the system dictate that the fuel/air ratio should be altered, no provision exists for automatic correction, and the right combustion conditions can only be restored by manual intervention. To improve performance and safety, some form of automatic recognition and correction of these factors would be preferable.

If the fuel/air ratio is incorrect, combustion of the fuel will be affected and the results will be observable in the flue gases. This indicates that an effective way of optimising the combustion process is to change the fuel/air ratio automatically in response to measurements of the flue-gas content.

For all fossil-fuelled boilers, the oxygen content of the flue gases increases as the excess-air quantity is increased, while the carbon dioxide and water content decreases. The carbon monoxide content of the boiler's flue gases is a direct indication of the completeness of the combustion process and systems based on the measurement of this parameter have long been recognised as an effective mechanism for improving combustion performance in coal and oil-fired boiler plant [1]. However, experience indicates that the use of this gas as a controlling parameter is less advantageous in boilers fired on natural gas [2].

Measurement of the flue-gas oxygen content often provides a good indication of combustion performance, but it must be appreciated that the presence of 'tramp air' due to leakages into the combustion chamber can lead to anomalous readings. In the presence of significant leakage, reducing the air/fuel ratio to minimise the flue-gas oxygen content can result in the burners being starved of air. This is an area where systems based on carbon monoxide measurements provide better results since the carbon monoxide content of the gases is a direct indication of combustion performance and is unaffected by the presence of tramp air.

A system which adjusts the fuel/air ratio in relation to the flue-gas oxygen content is shown in Figure 5.5. The oxygen measurement is fed to a controller (5) whose output adjusts the fuel/air ratio by varying the multiplying factor of a gain block (8).

The transmitters used for measuring flue-gas oxygen are usually based on the use of zirconium probes, whose conductivity is affected by the oxygen content of the atmosphere in which they are installed. True two-wire 4–20 mA analysers are now available (Figures 5.6 and 5.7), and are both accurate and reliable.

The flue gases leave the combustion chamber through ducts of considerable cross-sectional area and it is inevitable that a significant degree of stratification will occur in the gases as they flow to the chimney. Air entering the furnace through the registers of idle burners will tend to produce a higher oxygen content in the gases flowing along one area of the duct than will be present in another area, where fewer burners may be idle.

It is therefore necessary to take considerable care that any gas analysis provides a truly representative sample of the average oxygen content, and this demands that great care should be exercised over the selection of the

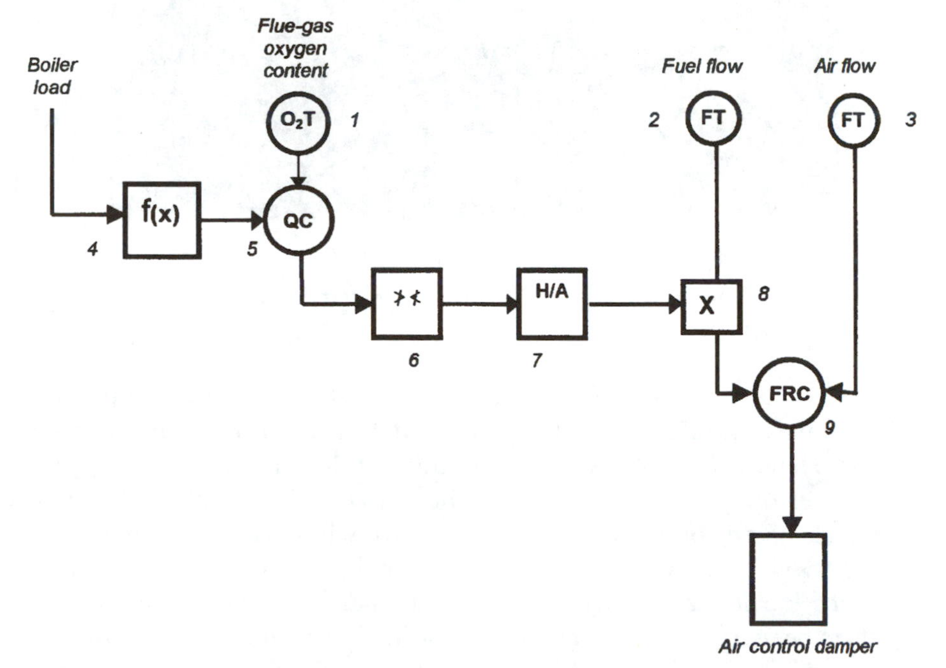

Figure 5.5 Oxygen trimming of fuel/air ratio

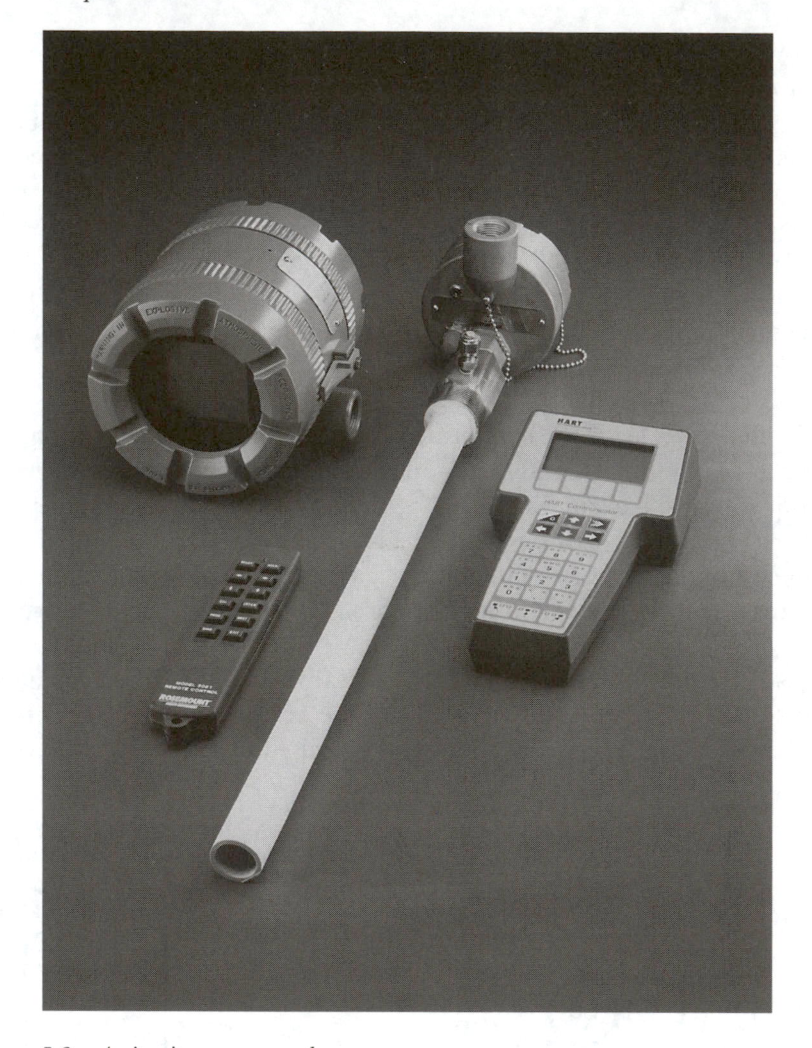

Figure 5.6 *An in-situ oxygen analyser*
© Fisher Rosemount Ltd. Reproduced by permission

location of the analyser. With larger ducts it may be necessary to provide several analysers. The signals from these can be combined, or the operator can be given the facility to select one or more of them for use.

A better option is now available. The power and flexibility of modern computer-based control systems allows for truly intelligent sampling to be applied, where the system recognises the dynamic status of the plant, such as which burners are being fired, and automatically selects the analyser signal to be used, or intelligently mixes the analyser signals to optimise performance. The installation of such a system requires careful observation of

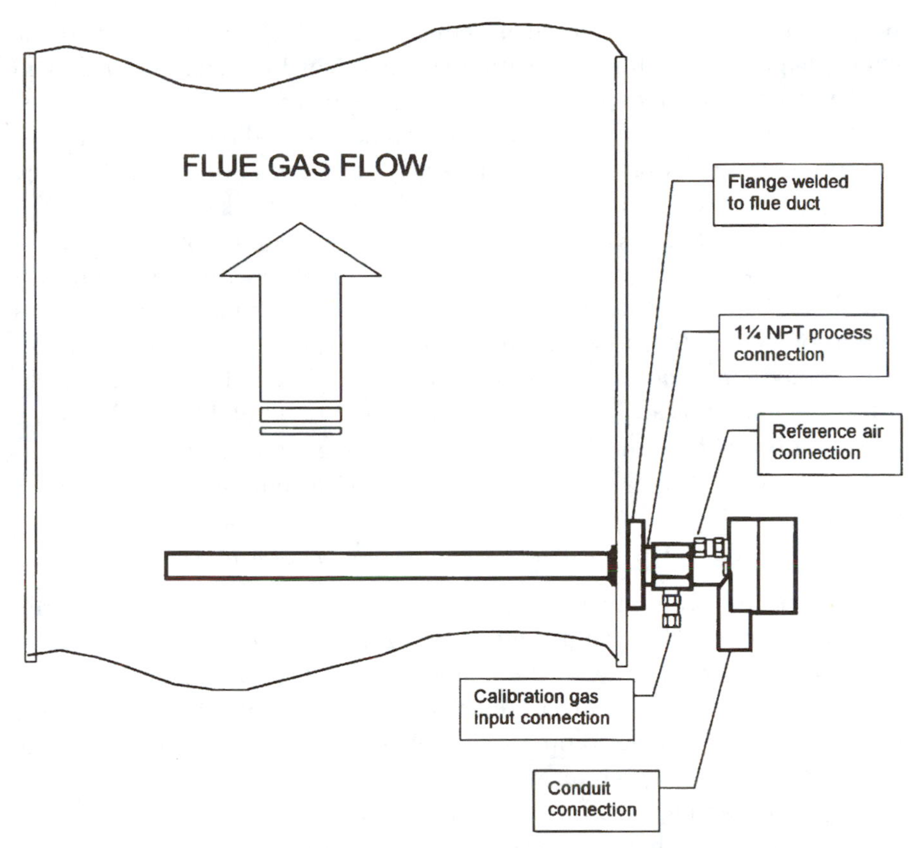

Figure 5.7 Installation of an in-situ oxygen analyser

the plant performance over an extended period and the development and subsequent application of a suitable system based on those observations.

Although such techniques are possible. Despite the considerable advances that have been made in gas-analyser technology over the past few years, fuel/air ratio trimming on the basis of gas analysis is still treated with some reservation. It is generally accepted that the measurements may occasionally fail or be misleading and for this reason it is usual to allow manual intervention in the absence of reliable oxygen control. In Figure 5.5 this facility is provided by the hand/auto station (7). In addition, a maximum/minimum limiter block (6) restricts the amount of adjustment that is permitted, to constrain the effects of anomalous or invalid measurements or incorrect control actions.

This system also characterises the set-value signal for the oxygen controller over the boiler's load range by means of a function block (4), providing for higher excess-air operation at low loads. The indication of boiler load may be obtained from either steam flow or air flow, and the exact shape and parameters of the oxygen versus load characteristic will be defined by the boiler designer or process engineer.

In practice, facilities may also be incorporated to allow the operator to adjust the system by biassing the load signal upwards or downwards at any given point to yield better combustion with reduced stack emissions.

Because the oxygen content of air is 21% by volume (or roughly 23% by weight), a given change in oxygen content represents approximately five times that change in terms of excess air. Since it is indeed *air flow* that is being controlled, the oxygen loop must recognise the presence of this high-gain component, and the gain of the controller (5) should be set at a kick-off low value (typically 0.25, or a proportional band of 400%). The time constants of the fuel/air/flue-gas system are long, and the integral term of the oxygen controller will therefore also tend to be long.

5.1.3.2 Combining oxygen measurement with other parameters

The use of an oxygen-trim signal on its own can be misleading, for the reasons noted earlier, and better performance can be obtained by combining oxygen trim with the opacity of the flue gases, since reducing the air flow eventually results in the production of visible smoke. However, it is usually undesirable to operate a boiler in the region where smoke is being produced, and an improvement is to adjust the air flow on the basis of another parameter, such as carbon monoxide.

Figure 5.8 shows how the carbon monoxide and oxygen measurements can be combined to trim the fuel/air ratio. Basically, the system comprises two gas-analysis controllers (6 and 10) whose set-value signals are determined in relation to the boiler load (via function generators 5 and 7). However, the set value for the oxygen controller is also trimmed by the output of the carbon monoxide controller (the two signals being combined in summator 9). Hand/auto facilities enable the system to operate with both analysers in command, or with only oxygen trim in service (the CO controller being on manual at hand/auto station 8), or with fully manual fuel/air ratio adjustment (hand/auto station 12 being on manual, to isolate both gas-analysis controllers).

In another variant of this system, either of the two flue-gas analysis controllers can be selected for operation, either by manual intervention or automatically by means of a maximum-selection function.

5.1.3.3 Using carbon-in-ash measurements

In boilers burning solid fuels, the carbon content of the ash has traditionally been used to provide an indication of the completeness of combustion,

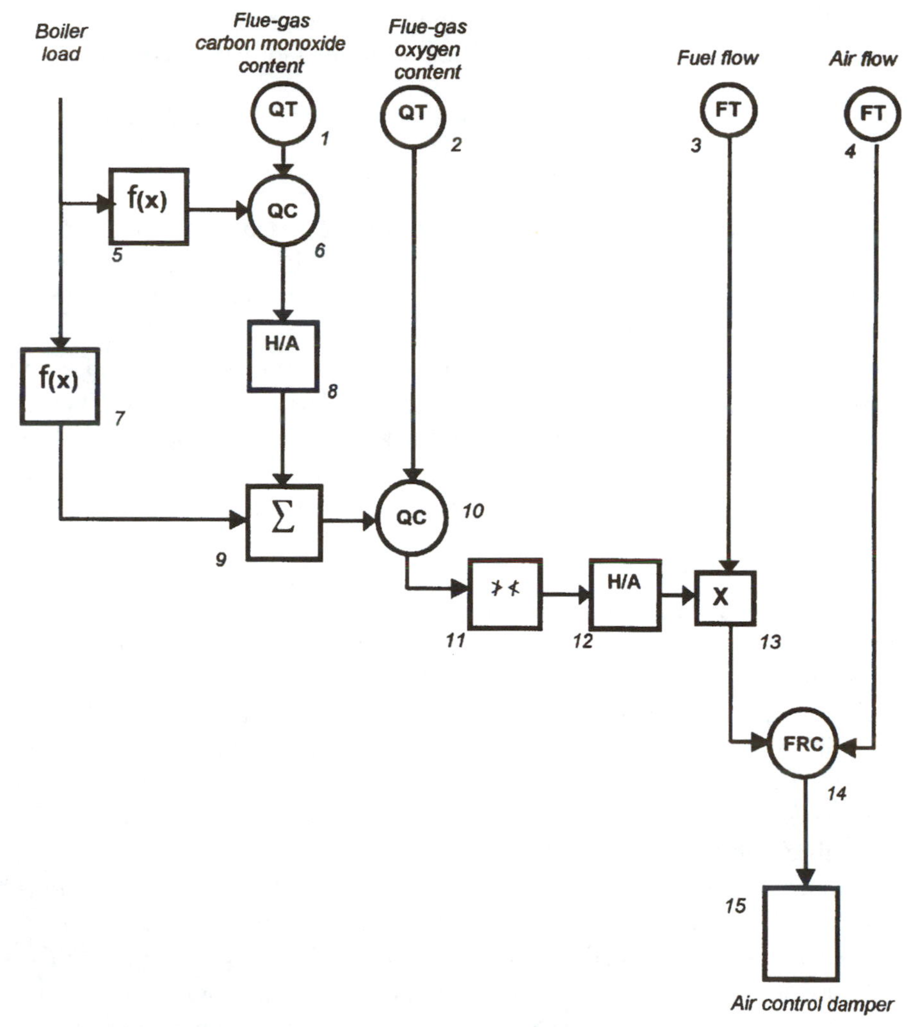

Figure 5.8 Combined CO and O_2 trimming of fuel/air ratio

since any carbon remaining in the ash indicates that incomplete combustion has occurred. Until comparatively recently, accurate online carbon-in-ash sampling was not possible and measurement of this parameter required manual sampling and analysis. With the emergence of online analysers the picture has changed, and tests have indicated that online measurement can play a useful part in optimising the combustion process [3]. In addition, analysis of unburned carbon can indicate whether the coal mills (pulverisers) require adjustment. However, the long transfer-time constants of the combustion process coupled with the comparatively slow response of the instruments and problems of stratification [4] suggest that this technique is only useful for long-term correction of firing, where relatively stable load conditions can be maintained for extended periods.

5.1.4 Multiple-burner systems

The systems that have been described so far are based on the adjustment of the total quantity of fuel and air that is admitted to the combustion chamber. This approach may suffice with smaller boilers, where adjustment of a single fuel valve and air damper is reasonable, but larger units will have a multiplicity of burners, fuel systems, fans, dampers and combustion-air supplies. In such cases proper consideration has to be given to the distribution of air and fuel to each burner or, if this is not practical, to small groups of burners. Again, suitable standards have been developed by the NFPA for the design of the plant and control systems of such boilers [5].

The concept of individually controlling air registers to provide the correct fuel/air ratio to each burner of a multiburner boiler has been implemented, but in most practical situations the expense of the instrumentation cannot be justified. Oil and gas burners can be operated by maintaining a defined relationship between the fuel pressure and the differential pressure across the burner air register (rather than proper flow measurements), but even with such economies the capital costs are high and the payback low. The need to provide a modulating actuator for each air register adds further cost.

A more practical option is to control the ratio of fuel and air that flows to groups of burners. Figure 5.9 shows how the principles of a simple cross-limited system are applied to a multiburner oil-fired boiler. The plant in this case comprises several rows of burners, and the flow of fuel oil to each row is controlled by means of a single valve. The combustion air is supplied through a common windbox, and the flow to the firing burners is controlled by a single set of secondary-air dampers.

In most respects the arrangement closely resembles the basic cross-limited system shown in Figure 5.4, with the oil flow inferred from the oil pressure at the row. A function generator is used to convert the pressure signal to a flow-per-burner signal, which is then multiplied by a signal representing the number of burners firing in that row, to yield a signal representing the total amount of oil flowing to the burners in the group.

The system operates in exactly the same way as the basic configuration of Figure 5.4, and it is repeated for each row of burners, so that the ratio of total fuel-oil flow to total air flow entering the boiler is maintained at the desired value. The master demand and the oxygen-trim signals are fed to all the rows to keep the firing rate in step with the load demand and the flue-gas oxygen content at the correct level.

This basic configuration is not restricted to oil-fired boilers. It can also be used with gas-fired plant and it can be applied to systems burning a mixture of fuels, with suitable modifications as will now be described.

5.2 Working with multiple fuels

The control systems of boilers burning several different types of fuel have to recognise the heat-input contribution being made at any time by each of the fuels, and the arrangements become more complicated for every additional fuel that is to be considered.

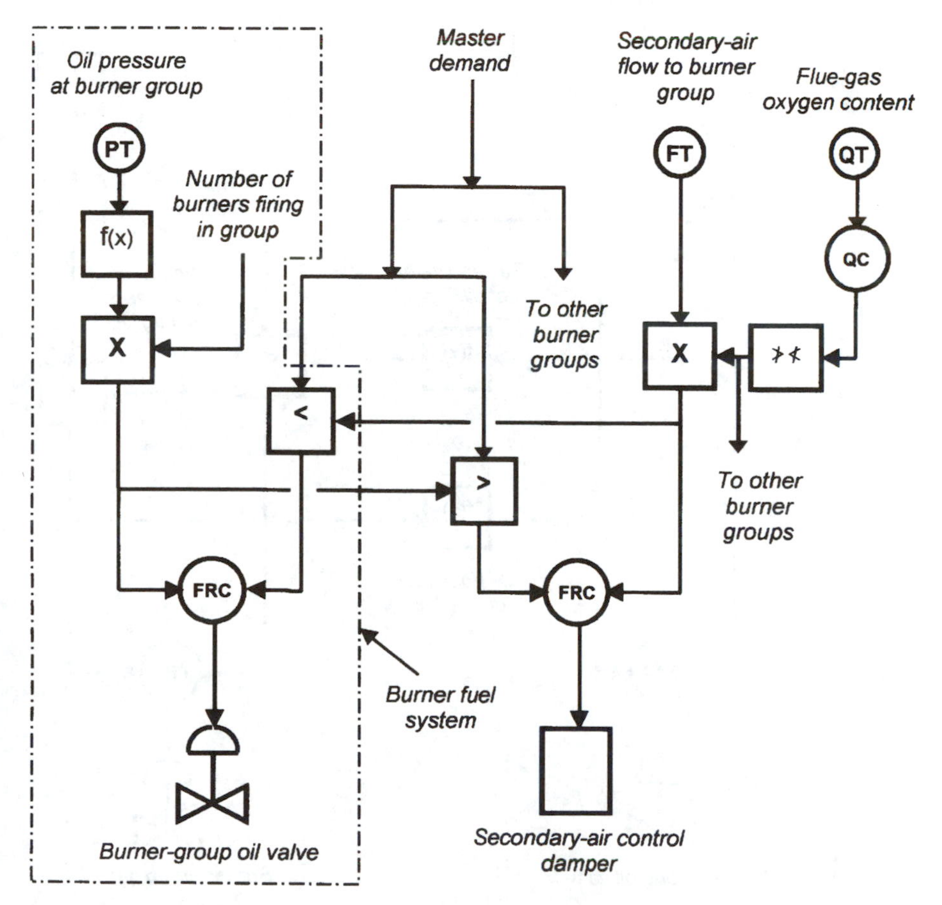

Figure 5.9 A control system for multiple burners (one burner group shown)

Figure 5.10 shows a system for a boiler burning oil and gas. The similarities to the simple cross-limited system are very apparent, as are the commonalities with the fuel-control part of the multiburner system (shown within the chain-dotted area of Figure 5.9).

The cross-limiting function is performed at the minimum-selector block (5) which continuously compares the master demand with the quantity of combustion air flowing to the common windbox of the burner group. The gain block (6) translates the air flow into a signal representing the amount of fuel whose combustion can be supported by the available secondary air.

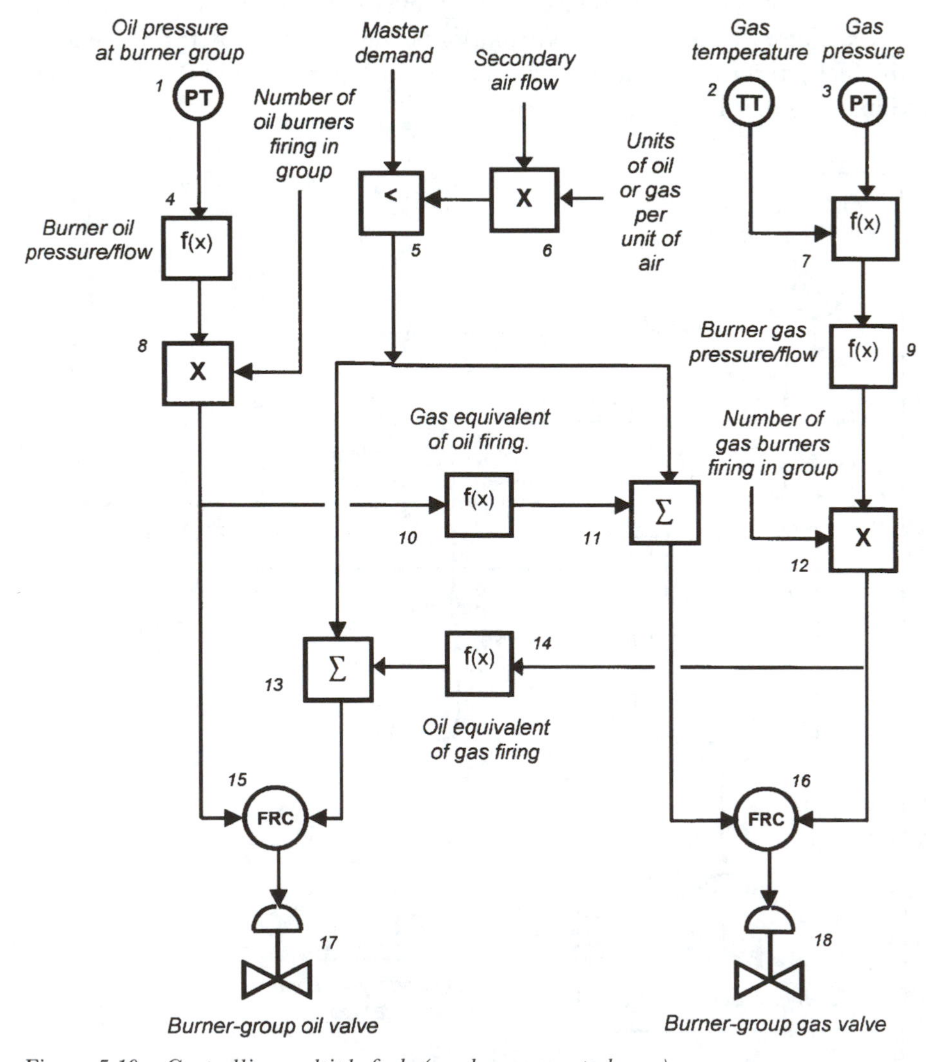

Figure 5.10 Controlling multiple fuels (one burner group shown)

The selected signal (the load demand or the available air) ultimately forms the desired value of both the gas and oil closed-loop controllers. But, before it reaches the relevant controller a value is subtracted from it, which represents the heat contributed by the other fuel (converted to the same heat/m^3 value as the fuel being controlled). The conversion of oil flow to equivalent gas flow is performed in a function generator (10), while the other conversion is performed in another such block (14). Each of the two summator units (11 and 13) algebraically subtracts the 'other-fuel' signal from the demand.

Note that, in the case of this system, the gas pressure signal is compensated against temperature variations, since the pressure/flow relationship of the gas is temperature-dependent.

As before, each fuel-flow signal represents the flow *per burner* and so it has to be multiplied by the number of burners in service in order to represent the total fuel flow.

These diagrams are highly simplified, and in practice it is necessary to incorporate various features such as interlocks to prevent overfiring and to isolate one or other of the pressure signals when no burner is firing that fuel. (This is because a pressure signal will exist even when no firing is taking place.)

5.3 The control of coal mills

So far, we have looked at boilers where the input of fuel can be measured and where its flow can be regulated by means of one or more valves. With boilers burning coal, the mill (or pulveriser) system must be taken into consideration. The mills have already been described in Chapter 3, now we shall look at how they are controlled.

But first it has to be understood that, because the mill has to meet defined performance guarantees, the control strategy to be applied in a given installation must be developed in association with the manufacturer of the mill. Once that strategy has been agreed it must be applied to each of the mills that feed the boiler. The demand is fed in parallel to all the mill sub-systems, with facilities for biassing the signal to any one of them with respect to the others.

5.3.1 The 'load line'

The drop in pressure experienced by air flowing through a mill will be determined by the geometry of the mill, the amount of coal in it and the volume of air flowing through it. Figure 5.11 shows schematically that a

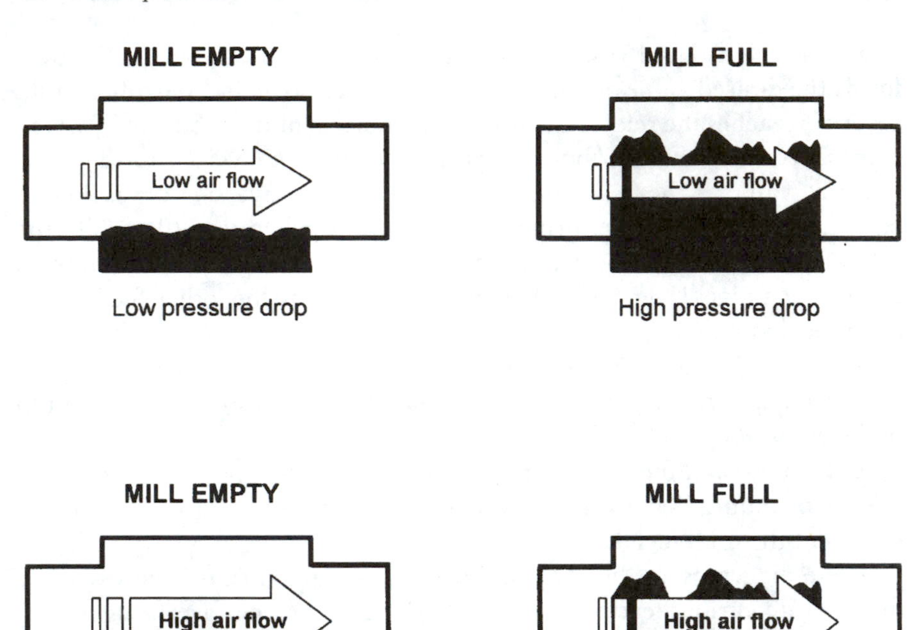

Figure 5.11 Effect of coal load and air flow on cross-mill differential pressure

high pressure-drop across the mill may be the result of a high coal load in the mill or a high air flow through it, or a combination of both. The air-flow rate will bear a square-law relationship to the differential pressure across the mill, and the differential pressure across a restriction such as a flow nozzle or an orifice plate will also have a square-law relationship with the air flow. From this, it can be appreciated that the characteristic curve relating the mill differential pressure and the primary-air differential pressure will be a straight line. This is called the 'load line' and is specific to a given design of mill operating under defined conditions. The manufacturer will define the correct load-line parameters and scales for a given design of mill.

5.3.1.1 Load control strategies for pressurised mills

With pressurised mills, some control systems operate on the principle of comparing the two differential-pressure signals and modulating the feeder speed to keep the relationship between the two in track with the load line, as shown in Figure 5.12. The methods of varying the speed of the feeder include variable-ratio gearboxes or variable-speed motors.

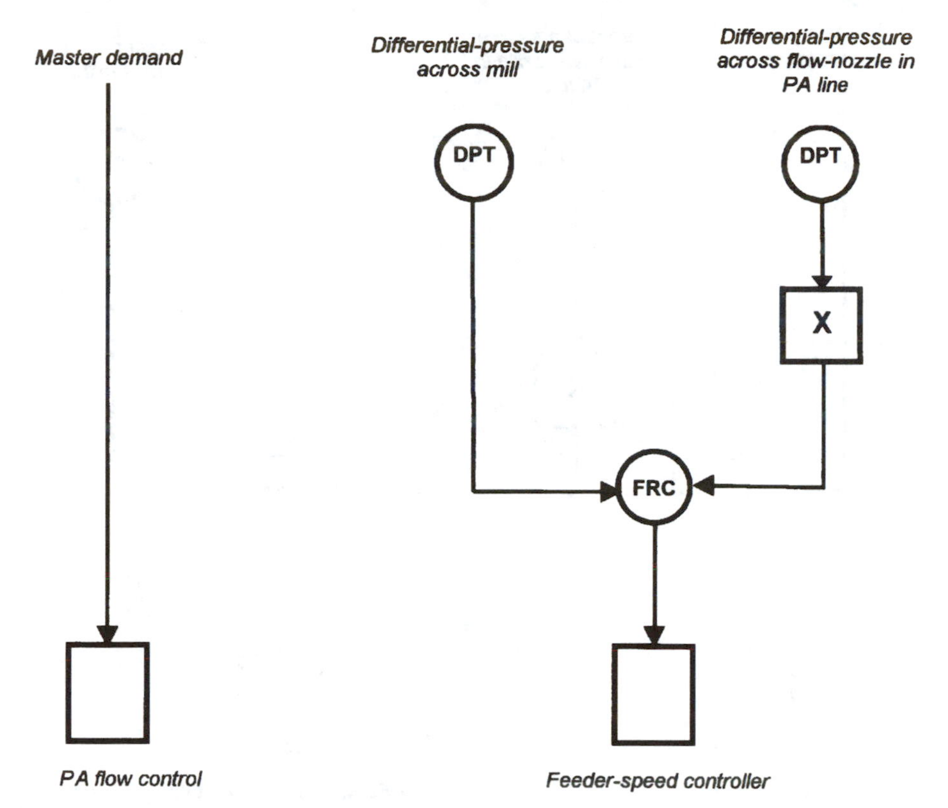

Figure 5.12 Mill-differential/PA-differential control system

The speed of the feeder is sometimes fed back to the master system as an indication of coal flow, to provide a degree of closed-loop operation. It is not a perfect solution, since a change in the calorific value of the coal cannot be determined by this system. But, in the absence of reliable and fast systems for measuring the heat input from coal, it is as good as can be achieved.

Although the system described above provides an adequate method of control, it cannot deal with changes in the primary-air (PA) flow caused by external factors. Therefore, if the PA flow changes, the system must wait for the resulting change in steam pressure before a correction can be made.

An approach to overcoming this limitation is to provide closed-loop control of the primary-air flow, as shown in Figure 5.13. Here, because the system detects and immediately reacts to changes in PA flow, and adjusts the flow-control damper to compensate, disturbances to steam production are minimised. Again, a feeder-speed signal, representing fuel flow, is fed

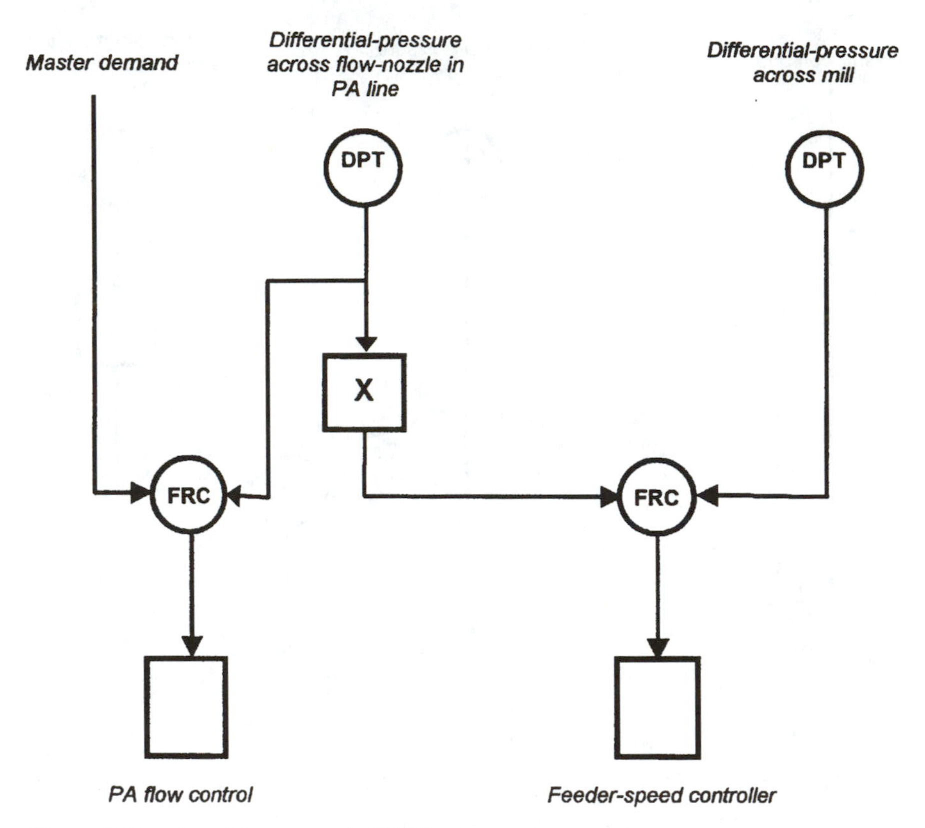

Figure 5.13 Closed-loop control of PA flow

back to the master system to provide closed-loop correction of speed changes, which would otherwise introduce disturbances to the steam pressure.

Both of these systems adjust the feeder speed *after* the PA flow has been changed, and this can lead to delayed response to changes in demand. Figure 5.14 shows a system that adjusts the feeder speed *in parallel with* the PA flow. This also shows some practical refinements: a minimum-limit block that prevents the PA flow from being reduced below a predetermined limit, and a minimum selector block which prevents the coal feed being increased above the availability of primary air (the bias unit sets the margin of air over coal).

5.3.1.2 Load control systems for suction mills

In broad terms, the load-control strategies for suction mills follow similar principles to those of the pressurised mills as described above. A very

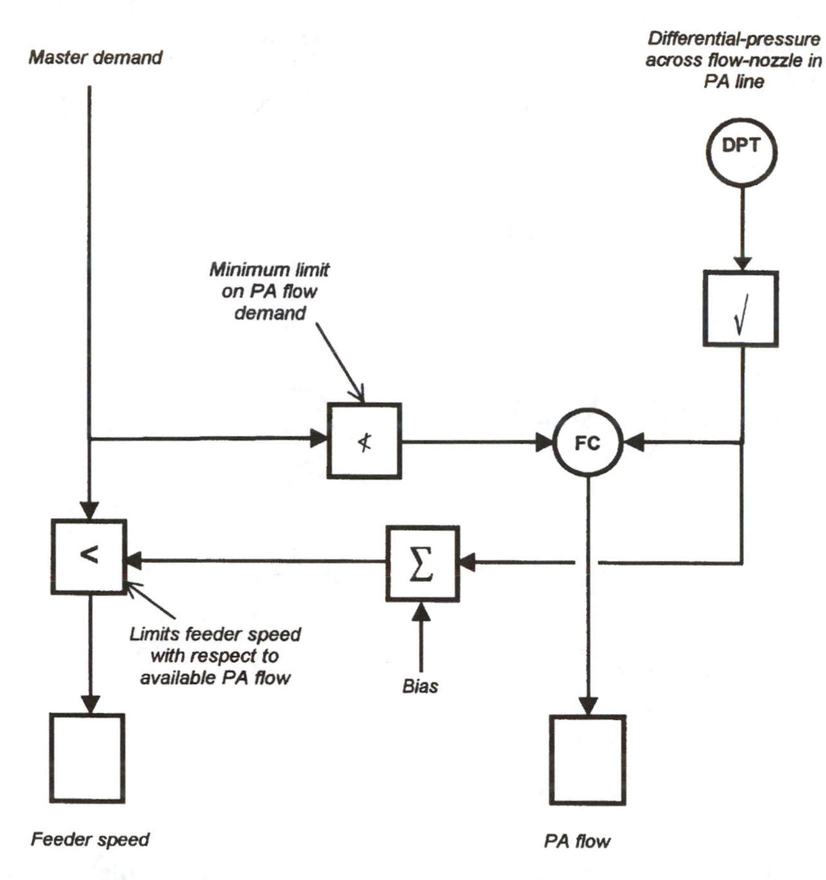

Figure 5.14 *Parallel control of feeder speed and PA flow*

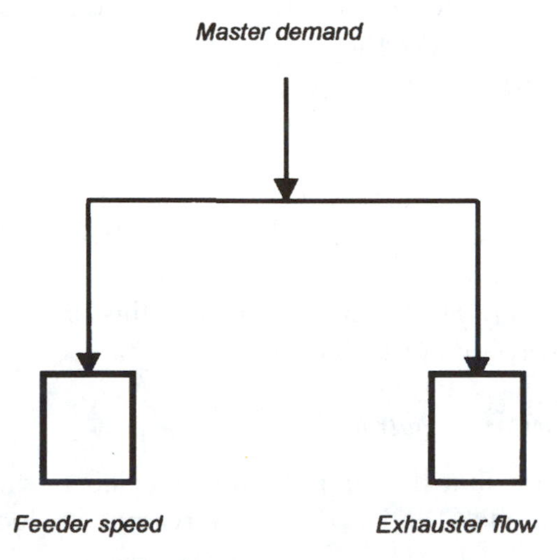

Figure 5.15 *Simple suction-mill control*

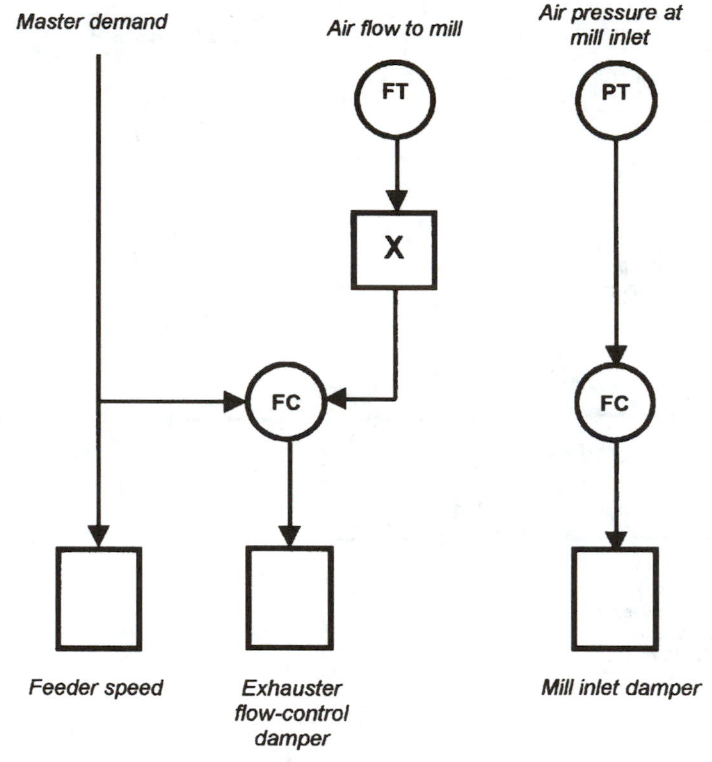

Figure 5.16 Improved control of suction mill

simple technique is to adjust the speed of the coal feeder in parallel with the flow through the exhauster, as shown in Figure 5.15. Here again, the feeder speed is returned to the master system to correct for speed variations that would otherwise disturb the steam pressure.

This system provides open-loop operation of the mill and, once again, improved performance can be achieved by the use of a closed loop around the air flowing through the mill, as shown in Figure 5.16. In this system, an additional control loop maintains a constant air pressure at the mill inlet.

With these systems, it is again necessary to feed back to the master system a signal that represents the input of fuel from the mill to the combustion chamber. Feeder speed provides this function, and thereby minimises steam-pressure disturbances.

5.3.2 Mill temperature control

It is very important that the temperature of the air in the mill should be maintained within close limits. For many reasons, including inadequate drying of the coal, combustion efficiency will be reduced if the temperature is too low, while too high a temperature can result in fires or explosions occurring in the mill. The control techniques for both pressurised and suction mills involve mixing hot and cold air streams to achieve the correct temperature. However, whereas pressurised coal mills require the use of two dampers for this purpose (one controlling the flow of hot air, the other the cold air) in a suction mill only one damper needs to be adjusted, to admit more or less cold air into the stream of hot air being drawn into the mill by the exhausters.

Figure 5.17 shows a temperature control system for a pressurised mill, with one actuator provided for the hot-air damper and another for the tempering-air damper.

Sometimes the two dampers are linked mechanically and positioned by a single actuator. The use of two separate actuators adds cost, but allows for a greater degree of operational flexibility since it allows the opening of each damper to be biassed with respect to the other from the central

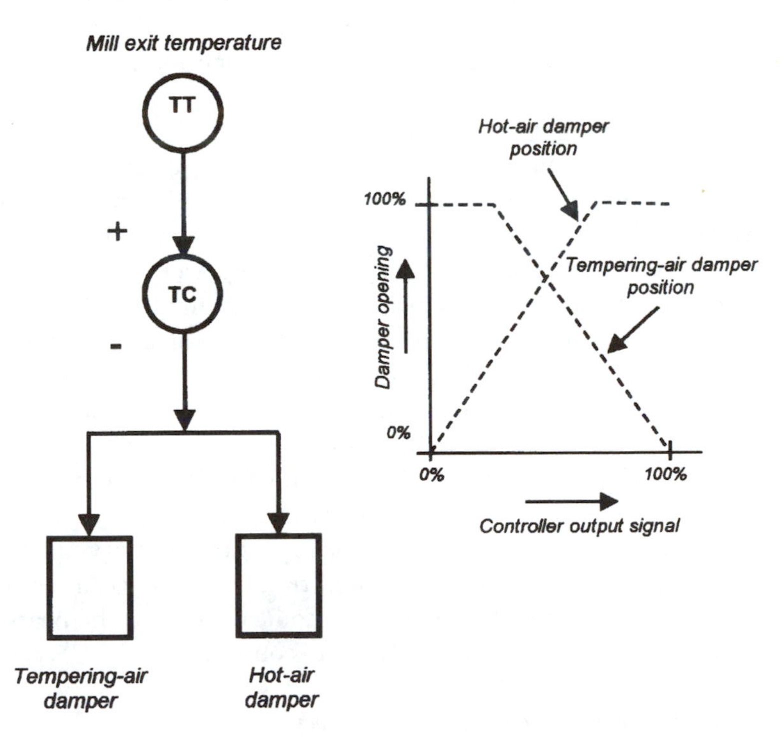

Figure 5.17 Mill temperature control

control room. This enables the operator (or a sophisticated control system) to optimise mill performance whilst still maintaining the mill temperature at the correct value.

5.3.3 Controlling multiple mills and multiple fuels

Large coal-fired boilers are provided with several coal mills, each of which has its own control subsystem as described above, and in addition they invariably burn other fuels as well as coal.

Figure 5.18 shows the mill-control system of such a plant in simplified form. It is presented here to illustrate a requirement that is an integral part of boiler control systems: the need to handle applications where a single controller sends commands to several subloops in parallel, and where any of the subloops may be isolated at will from the controller.

Here, each mill feeds a group of burners (say six), and each of these groups may also fire fuel-oil. Since, at any given time, any mill group may be out of service, operating at a fixed throughput, or otherwise requiring independence from the other groups, the overall loop gain will change, and this is addressed by the gain-compensation block (item 4) in the master-demand signal line. The demand signal from this block is fed to each group via individual hand/auto stations, one for each mill group (item 10).

The output of each of these stations eventually becomes the desired value for the relevant primary-air flow controller (17), but first the heat contribution from any oil burners firing in that group must be taken into consideration. This input is derived from a measurement of the oil pressure at the burners in the group (1), converted to represent the oil flow per burner (by means of function block 2) and then multiplied (4) by the number of oil burners in that group that are firing at the time. The resulting signal is then converted (9) to represent the amount of coal that would equate to that quantity of oil, and this is subtracted from the master demand (block 12) to represent the amount of coal firing that is needed from the group. This firing demand is prevented from falling below a safe predetermined value (minimum-limit block 15).

By accounting for the oil firing, the opening of the primary-air damper is immediately adjusted if an oil burner trips, or if one is brought into service, to compensate for the change, without waiting for the heat-input effects to be detected via the master-pressure controller.

5.3.3.1 The challenge of hand/auto changeover

The heat input from a large coal mill can be as much as 100 MW, but the mechanical design of the mill and its auxiliaries is such that it can vary

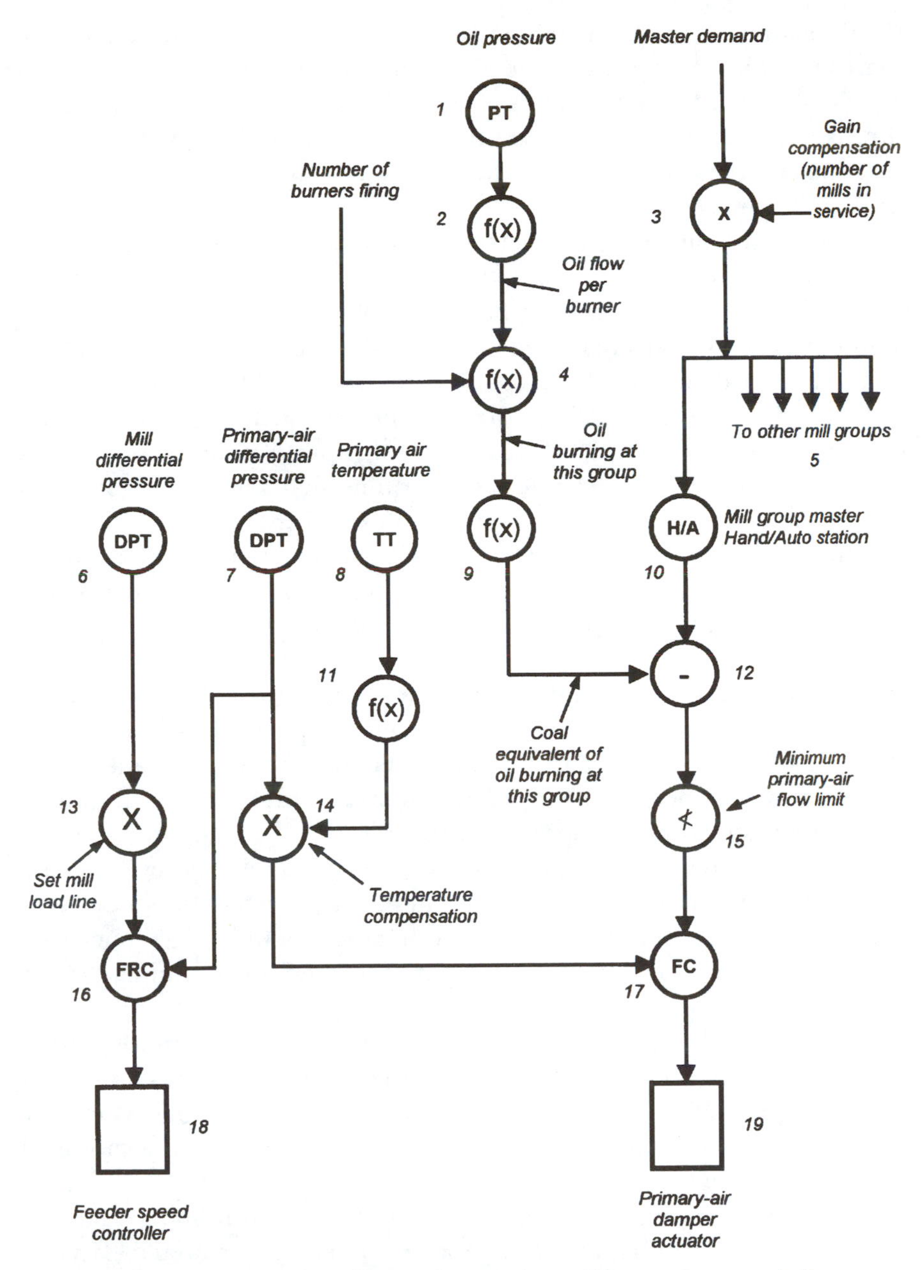

Figure 5.18 A comprehensive mill control system (one mill group shown, excluding temperature control)

the throughput by only a comparatively small amount, certainly no more than 50%. Therefore, the introduction of one mill to the heat input of such a boiler amounts to a step change of as much as 50 MW, and the change in throughput that can be smoothly modulated is also 50 MW. Such large step-changes require efficient modulation of any other fuels that are being fired at the same time.

These factors make it impractical to consider starting up more than one mill at a time and require the facility of allowing any mill to be operated under manual or automatic control, independent of the others. This brings about a severe challenge to the DCS software.

The master demand is fed in parallel to several subloops, one for each mill group. On start-up of the plant all of these will be under manual control. When the mill has reached a throughput of roughly 50% of its capacity, or when other conditions determine that automatic control is now possible, the operator will switch the master demand into service. The difficulty is that up to that instant, the system cannot be made aware of which mill group is about to be transferred to respond to the master signal, and each group may be operating at a very different throughput from any other.

While a loop is being transferred from manual to automatic control (or vice versa), it is important that the plant is not subjected to a sudden disturbance. At the moment of changeover, the 'hand' and 'automatic' signals must be equal. This is called 'bumpless transfer', and it can be achieved by providing the operator with indications of both signals so that they can be made equal before changeover is initiated. However, such a system would not be acceptable in most cases, since the process of changing from one mode of control to another should be as quick and simple as possible, and should not require the operator to unduly disturb the operation of the plant.

To achieve what is know as 'procedureless, bumpless transfer' from manual to automatic control, a common technique is to make the controller output follow (or 'track') the manual demand, so that when the system is switched to automatic the signal to the actuator is not subjected to a sudden change.

This is easy enough with a single controller positioning a single actuator, but what happens when one controller commands several subloops as shown in Figure 5.18? It is clearly impossible to force the master controller output to adopt a value that cannot be known ahead of time, or to change the output of the controller if it is already modulating one or more mills.

This problem is frequently not recognised by DCS vendors who have little or no experience of boiler control, and it can be quite difficult to explain it to them. But understanding it and resolving it are absolutely essential if the system is to be expected to operate smoothly and with minimal operator intervention. Various solutions have been developed, such as 'freezing' the master demand while the transfer is effected and gradually ramping one signal up or down to match the other. It is important, however, that the DCS vendor should be able to demonstrate the solution offered within their system, and that they should be able to demonstrate its use on an existing power plant.

5.3.3.2 Complexity of screen displays

In considering the operator displays associated with a system such as that shown in Figure 5.18, attention should be given to the vast amount of information that must be provided. The diagram given here is necessarily simplified, and excludes the many interlocks and other functions that are required in reality. When a practical plant is considered it soon becomes apparent that accommodating the amount of information and control facilities can lead to very cluttered display screens.

Clearly, the mill groups are carbon copies of each other, varying only in respect to the tag numbers of each item and the dynamic information relating to each area of the plant. It is therefore reasonable to display only one group at a time on the screen, allowing it to be started, adjusted or stopped as required. However, to avoid making any mistakes, the operator should be very clearly and unambiguously informed of which group is displayed at any time. Also, a master display should enable the operator to view the status of the entire set of mills feeding the boiler.

The development of these operator displays is therefore unusually demanding and if insufficient time or money is allocated to the performance of this task the results can be at best unwieldy and at worst dangerous.

5.4 Draught control

In Chapter 3 we saw that, in a fired boiler, the air required for combustion is provided by one or more fans and the exhaust gases are drawn out of the combustion chamber by an additional fan or set of fans. On boilers with retro-fitted flue-gas desulphurisation plant, additional booster fans may also be provided. The control of all these fans must ensure that an adequate supply of air is available for the combustion of the fuel and that the combustion chamber operates at the pressure determined by the boiler designer. In a fluidised-bed boiler the air must also provide the pressure required to maintain the bed in a fluid state.

All of the fans also have to contribute to the provision of another important function—purging of the furnace in all conditions when a collection of unburned fuel or combustible gases could otherwise be accidentally ignited. Such operations are required prior to light-off of the first burner when the boiler is being started, or after a trip.

The control systems for the fans have to be designed to meet the requirements of start-up, normal operation and shut-down, and to do so in the most efficient manner possible, because the fans may be physically large and require a large amount of power for their operation (several MW in some cases). In addition, as we saw in Chapter 3, the performance constraints of the fans, such as surge and stall, have to be recognised, if necessary by the provision of special control functions or interlocks.

Chapter 3 also described the methods of controlling the throughput of the fans, i.e. pitch-control, dampers, vanes or speed adjustment. In the present chapter we shall examine how these elements are adjusted to address the operational requirements of the boiler.

5.4.1 Maintaining the furnace draught

Apart from supplying air to support combustion, the FD fans have to operate in concert with the ID fans to maintain the furnace pressure at a certain value. The heavy solid line of Figure 5.19 shows the pressure profile through the various sections of a typical balanced-draught boiler system. It shows the pressure from the point where air is drawn in, to the point where the flue gases are exhausted to the chimney, and demonstrates how the combustion chamber operates at a slightly negative pressure, which is maintained by keeping the FD and ID fans in balance with each other.

If that balance is disturbed the results can be extremely serious. Such an imbalance can be brought about by the accidental closure of a damper or by the sudden loss of all flames. It can also be caused by maloperation of the FD and ID fans. The dashed line on the diagram shows the pressure profile under such a condition, which known as an 'implosion'. The results of an implosion are extremely serious because, even though the pressures involved may be small, the surfaces over which they are applied are very large and the forces exerted become enormous. Such an event would almost certainly result in major structural damage to the plant.

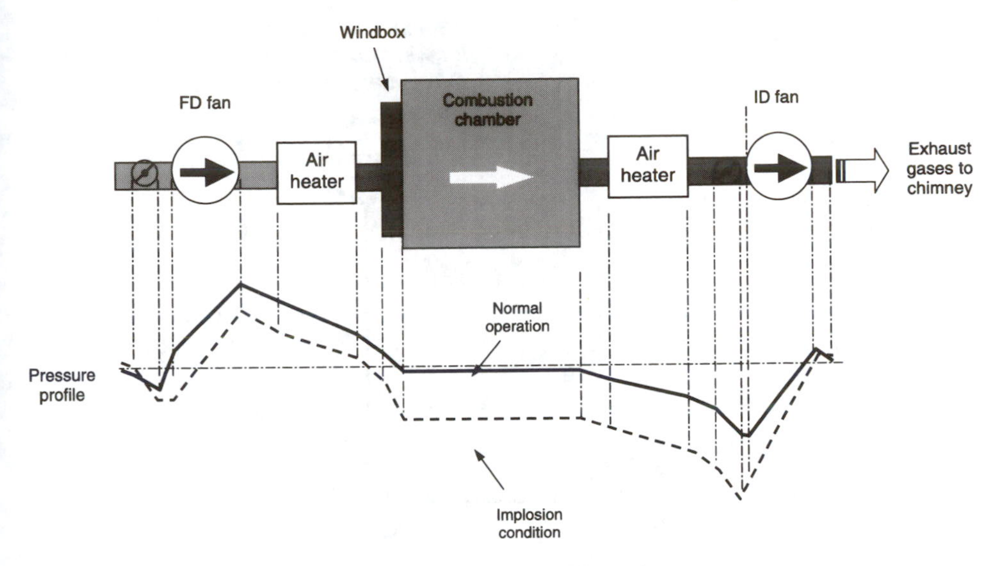

Figure 5.19 Draught profile of a boiler and its auxiliary plant

5.4.2 Fan control

The throughput of two fans operating together can be regulated by a common controller or by individual controllers for each fan. Although a single controller cannot ensure that each fan delivers the same flow as its partner, this configuration is much simpler to tune than the alternative, where the two controllers can interact with each other and make optimisation extremely difficult. Whichever option is used, the control system must be designed to provide sufficient air to support combustion.

In the simplest case, the fan or fans will be driven by a cross-limited system (see Figure 5.4), but with multiburner installations the flow must be controlled for each burner or group of burners. The system shown in Figure 5.9 shows how this is arranged by regulating the secondary air flow to each burner group. In such cases this air supply is drawn from a common windbox which is maintained at a pressure which may be fixed or varying with boiler throughput.

Figure 5.20 shows how such a control system can be implemented. The desired-value signal for the pressure controller is derived from steam flow, so that the pressure in the windbox will change over the boiler load range, to a characteristic that will be defined by the process engineer. The

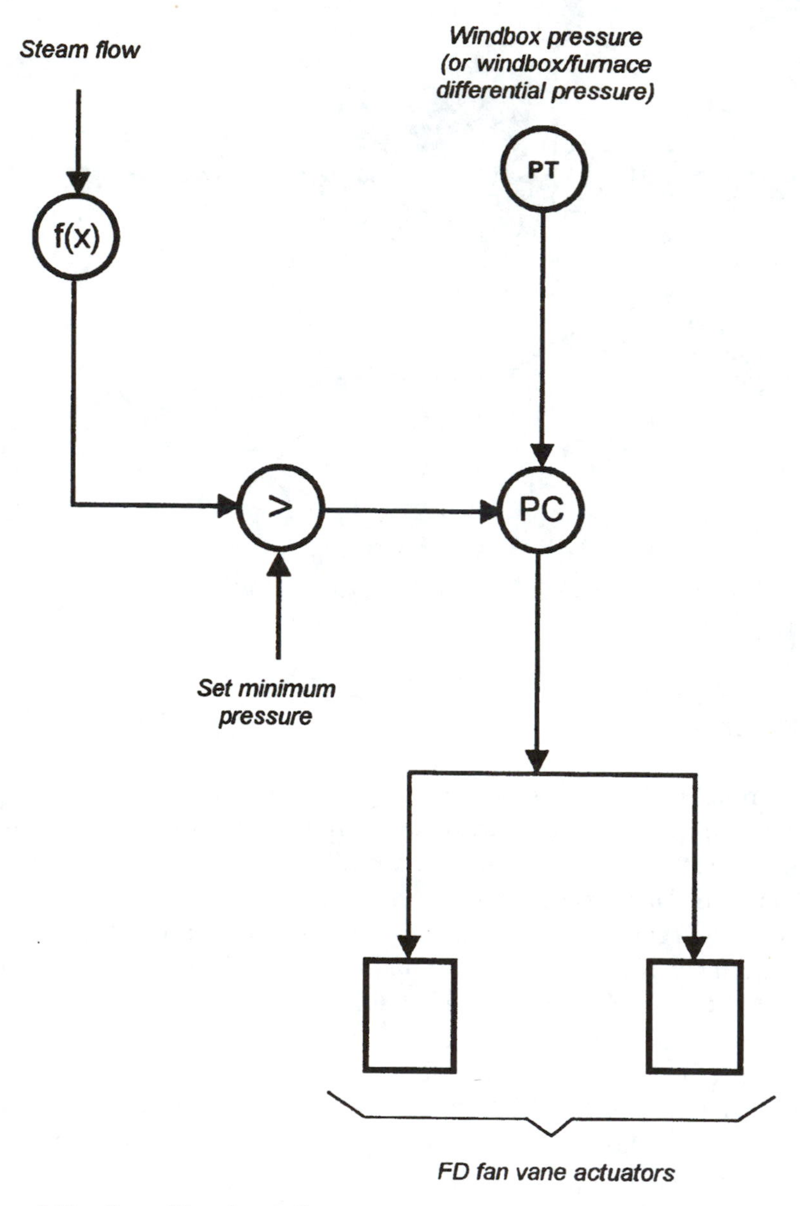

Figure 5.20 Controlling the windbox pressure

maximum-selector unit ensures that the pressure-demand signal cannot fall below a predetermined minimum value. The measured value for the controller can be based on a measurement of the windbox pressure or the windbox-to-furnace differential pressure (which is what the boiler designer would probably require).

5.5 Binary control of the combustion system

So far, we have considered only the modulating systems involved with the combustion plant. In practice, these systems have to operate in concert with binary control systems such as interlocks and sequences. The purpose of an interlock is to co-ordinate the operation of different, but interrelated plant items: tripping one set of fans if another set trips, and so on. The purpose of a sequence system is to provide automatic start-up or shutdown of the plant, or of some part of it.

The logic for interlock operations will be defined by the boiler designer and will probably have to comply with some local, national or international standard. The systems are very specific to the particular plant, and no attempt will therefore be made in this book to define these, because the objective here is to provide a general overview of boiler control systems.

However, one topic that we shall look at is burner management since, like modulating loops, this type of system is very dependent on the correct operation of input and output transducers.

5.5.1 Flame monitoring

The requirements for a comprehensive burner-management system (BMS) have already been discussed in Chapter 3, and attention was drawn there to the importance of flame monitoring.

Monitoring the status of a flame is not easy. The detector must be able to discriminate between the flame that it is meant to observe and any other in the vicinity, and between that flame and the hot surfaces within the furnace. The detector must also be able to provide reliable detection in the presence of the smoke and steam that may be swirling around the flame. To add to the problems, the detector will be required to operate in the hot and dirty environment of the burner front, and it will be subjected to additional heat radiated from the furnace into which it is looking.

With their attendant BMSs, flame scanners of a boiler are vital to the safety and protection of the plant. If insufficient attention is paid to their selection, or if they are badly installed or commissioned, or if their maintenance is neglected, the results can be, at best, annoying. The problems will include nuisance trips, protracted start-up of the boiler and the creation of hazardous conditions that could have serious safety implications.

Figure 5.21 shows a typical flame detector and the swivel-mounting that enables its sighting angle to be adjusted for optimum performance.

A flame scanner is a complex opto-electronic assembly, and modern scanners incorporate sophisticated technologies to improve flame recognition and discrimination. Although the electronics assembly will be

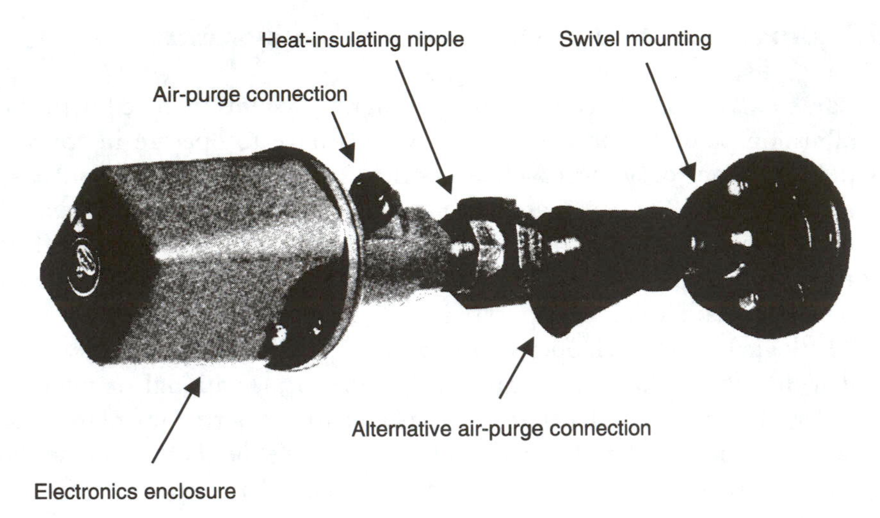

Figure 5.21 Typical flame scanner
© Fireye Ltd. Reproduced by permission

designed to operate at a high temperature (typically 65 °C), unless great care is taken this value could easily be exceeded and it is therefore important to take all possible precautions to reduce heat conduction and radiation onto the electronic components. The illustration shows how a heat-insulating nipple is used to prevent undue heat being conducted from the boiler structure to the electronics enclosure. It also shows two purge-air connections that are provided between the electronics enclosure and the swivel mount. Either of these connections may be used, the other being blanked off.

5.5.1.1 The requirements for purge air

The purge air that is supplied to the scanner serves two purposes: it provides a degree of cooling and it prevents dust, oil and soot from being deposited on the optical parts of the unit. The air should be available at each burner, even if the burner itself is not operating.

It should therefore be obvious that the air used for purging should be cool, dry and clean, and that it should be available at all times. But, in many cases these requirements are ignored, and the performance of the instrument is thereby inevitably degraded.

Purge air can be obtained from the instrument-air supply, or it can be provided by dedicated blowers. In some cases it is taken from the FD fan discharge. Each of these is viable, provided the requirements outlined above have been thoroughly considered. It is also important that the presence of the purge-air supply should be monitored and its loss transmitted to the DCS, because failure of the air supply could result in expensive and possibly irreparable damage to the scanners. Modern scanners include self-monitoring circuits that will warn of overheating. The scanner system should be fail-safe, as a failed system represents the loss of a critical link in the plant's safety chain. If it is overridden, the operator can become used to operating without it in place, and such lapses can eventually create a severe hazard.

Figure 5.22 shows an installation which clearly demonstrates examples of neglect, including a broken purge-air connection and a badly misaligned scanner. Unfortunately, in spite of the critical importance of reliable flame monitoring, it is not too difficult to find such examples on operating power plant.

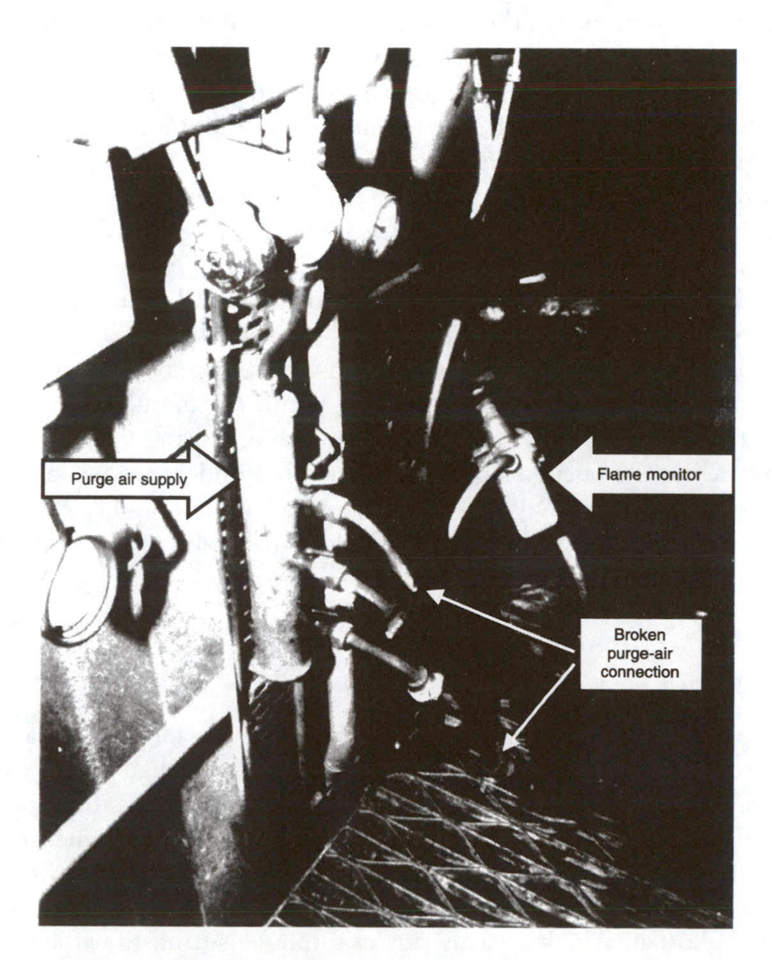

Figure 5.22 Example of a flame-scanner installation

5.5.1.2 Flame spectra

The spectrum of radiation from a flame is determined by many factors, including the type of fuel being burned and the design of the burner. The intensity of the flame tends to be low for gas and high for coal and oil. The flame will also flicker and, in general, low-NOx burners will demonstrate a lower flicker frequency than gun-type burners. Oil and coal flames tend to produce a higher degree of infrared radiation, whereas a gas flame is rich in ultraviolet radiation. Radiation in the visible part of the spectrum will also depend on these factors, but these days the tendency is to use detectors whose response is biassed towards either the infrared or the ultraviolet end of the spectrum, since emissions in these ranges provide better indication of a flame than visible radiation, which can be plentiful and misleading.

Each type of fuel also produces by-products of combustion, which affect the transparency of the flame and therefore the blanking effect it has on adjacent flames or on any flames on the opposite side of the furnace. Oil and coal flames tend to obscure infrared radiation, while gas flames produce water vapour which obscures ultraviolet radiation.

Table 5.1 shows one manufacturer's advice on the type of flame scanner to use in various applications. It is not intended that this table should be regarded as being absolute or rigorous. In certain circumstances a given type of flame scanner will provide better or worse performance than would appear to be indicated from the table. Reputable manufacturers will be pleased to provide application-specific guidance. At the design stage this advice will be based on previous experience of similar installations. For a retrofit on an existing plant, the manufacturer should be asked to carry out a comprehensive site survey, using various types of scanner, while the burners are started, operated under various loads, and stopped. Several tests may be required, and a survey may last for several days. The greater the attention that is paid to this study, the better will be the performance of the final installation.

5.5.1.3 Burner-management systems and plant safety

The design of the BMS will aim to address critical safety issues, and the sequences for a given type of boiler or burner will be defined in conjunction with the plant designer, bearing in mind the requirements of applicable codes such as NFPA 8502-95. In fact the NFPA standard defines in some detail the exact sequences involved in lighting-off, monitoring and running-down operations of burners, and shows how these are to be linked with the plant interlock systems (for example, ensuring that the furnace has been purged before any attempt can be made to initiate a burner light-

Table 5.1 Flame-scanner application guide
(© Fireye Ltd. Reproduced by permission)

Boiler type	Fuel type	Discrimination capability	
		Infrared	Ultraviolet
Front-fired	Gas	M	H
	Oil	H	H
	Coal	H	H
	Gas/oil	M	H
	Gas/coal	M	H
	Oil/coal	H	H
	Coal/oil/gas	M	H
Corner-fired	Gas	L	H
	Oil	H	H
	Coal	H	H
	Gas/oil	L	H
	Gas/coal	L	H
	Oil/coal	H	H
	Coal/oil/gas	L	H
Opposed-fired	Gas	L	H
	Oil	M	M
	Coal	M	M
	Gas/oil	L	M
	Gas/coal	L	M
	Oil/coal	L	M
	Coal/oil/gas	L	M

H = high, M = medium, L = low

off sequence). For these reasons, the sequences will not be described here. However, attention will be paid to certain safety-related aspects of BMSs.

Safety requirements are very comprehensively defined in every applicable standard. For example, NFPA 8502-95 describes the events and failures which should be recognised in the design of the system. The UK Health and Safety Executive (HSE) has described [6] in considerable detail the requirements for the safe design of a software-based system (defined as a programmable electronic system (PES)). However, in practice it is very difficult for the nonspecialist to determine whether or not a system is adequately fail-safe. Even using the checklists provided in the HSE document can be inadequate. For example, one item in the checklist asks:

Have adequate precautions been specified to protect against electrical interference in the environment of the PES with regard to:

(i) inherent design of the PES;
(ii) installation practices (e.g.: separation of power and signal cables);
(iii) Electromagnetic compatibility (EMC) test programme, including conducted interference on power supplies, electro-static discharges and radiated interference?

In that it is difficult to say what is or is not *adequate* in this context, this is a subjective assessment. For example, one system surveyed by the author found that the effectiveness of very comprehensive shielding in the DCS of a plant had been negated by the provision of poorly designed access doors.

One solution is to assume that a programmable system will occasionally generate incorrect commands and to therefore ensure that all its operations are continuously shadowed by another, independent, system. If a discrepancy occurs between the actions of the two systems a trip should be initiated or the relevant sequence prevented from being carried out.

5.6 Summary

Having looked at the control systems applying to the combustion and draught plant, in the next chapter we shall turn our attention to the feedwater systems.

5.7 References

1 ANSON, D, CLARKE, W. H. N., CUNNINGHAM, A. T. S. and TODD, P.: 'Carbon monoxide as a combustion control parameter', *J Inst Fuel*, 1971, **xliv** (363)
2 DUKELOW, S. G.: 'The control of boilers' (Instrument Society of America, Research Triangle Park, NC, USA, 1991)
3 ESKENAZI, D. V., D'AGOSTINI, M., LEVY, E. K. *et al.*: 'On-line measurement of unburned carbon', EPRI *Heat-rate Improvement* conference, EPRI, Palo Alto, CA, 1989
4 HURT, R. H., LUNDEN, M. M., BREHOB, E. G., and MALONEY, D. J.: 'Statistical kinetics for pulverized coal combustion'. Proceedings of the 26th international symposium on *Combustion*, Naples, Italy, 1996
5 NFPA 8502-95 'Standard for the prevention of furnace explosions/implosions in multiple burner boilers'. National Fire Protection Association, Batterymarch Park, Quincy, MA, USA

6 'Programmable electronic systems in safety related applications'. Health and Safety Executive, Library and Information Services, Broad Lane, Sheffield, S3 7HQ , UK

Chapter 6

Feed-water control and instrumentation

6.1 The principles of feed-water control

The objective of a feed-water control system may seem simple: it is to supply enough water to the boiler to match the evaporation rate. But as is so often the case with boilers, this turns out to be a surprisingly complex mission to accomplish. There are difficulties even in making the basic drum-level measurement on which the control system depends. The design of the control system is then further complicated by the many interactions that occur within the boiler system and by the fact that the effects of some of these interactions are greater or smaller at various points in the boiler's load range.

The control-system designer's task is to develop a scheme that provides adequate control under the widest practicable range of operational conditions, and to do so in a manner that is both safe and cost-effective. To do this it is necessary to understand the detailed mechanisms of the feed-water and steam systems and to be fully aware of the operational requirements.

In all but the smallest and simplest boilers, each of the interrelated factors has to be taken into account, and it is insufficient to rely on simple responses to the three parameters which seem to be relevant to the supply of feed water: steam flow, feed-water flow and the level of water in the drum.

6.2 One, two and three-element control

The level of water in the drum provides an immediate indication of the water contained by the boiler. If the mass flow of water into the system is greater than the mass flow of steam out of it, the level of water in the drum will rise. Conversely, if the steam output is greater than the feed inflow, the level will fall.

As stated in Chapter 2, the purpose of the drum is not only to separate the steam from the water but also to provide a storage reservoir that allows short-term imbalances between feed-water supply and steam production to be handled without risk to the plant. As the level of water in the drum rises, the risk increases of water being carried over into the steam circuits. The results of such 'carry-over' can be catastrophic: cool water impinging on hot pipework will cause extreme and localised stresses in the metal and, conversely, if the level of water falls there is a possibility of the boiler being damaged, partly because of the loss of essential cooling of the furnace water-walls.

Therefore, the target of the feed-water control system is to keep the level of water in the drum at approximately the midpoint of the vessel. Given this objective, it would appear that the simplest solution would appear to be to measure the level of water in the drum and to adjust the delivery of water to keep this at the desired value—feeding more water into the drum if the level is falling, and less if the level is rising. Unfortunately, the level of water is affected by transient changes of the pressure within the drum and the sense in which the level varies is not necessarily related to the sense in which the feed flow must be adjusted. In other words, it is not sufficient to assume that simply because the level is increasing the feed-water flow must be decreased, and vice versa.

This strange situation is due to effects known as 'swell' and 'shrinkage'. Boiling water comprises a turbulent mass of fluid containing many steam bubbles, and as the boiling rate increases the quantity of bubbles that is generated also increases. The mixture of water and bubbles resembles foam, and the volume it occupies is dictated both by the quantity of water and by the amount of the steam bubbles within it. If the pressure within the system is decreased, the saturation temperature is also lowered and the boiling rate therefore increases (because the temperature of the mixture is now higher in relation to the saturation temperature than it was before the pressure change occurred). As the boiling rate increases, the density of the water decreases, but since the mass of steam and water has not changed the decrease in density must be accompanied by an increase in the volume of the mixture.

By this mechanism the level of water in the drum appears to rise, a phenomenon referred to as 'swell'. The rise of level is misleading: it is not indicative of a real increase in the mass of water in the system, which would require the supply of water to be cut back to maintain the status quo. In fact, if the drop in pressure is the result of the steam demand suddenly increasing, the water supply will need to be increased to match the increased steam flow.

'Shrinkage' is the opposite of swell: it occurs when the pressure rises. The mechanism is exactly the same as that for swell, but in the reverse direction. Shrinkage causes the level of water in the drum to fall when the steam flow decreases, and once again the delivery of water to the boiler must be related to the actual need rather than to the possibly misleading indication provided by the drum-level transmitter.

If a slow change of steam flow occurs, all is well because the pressure within the system can be controlled. It is when rapid steam-flow changes happen that problems occur since, due to swell or shrinkage, the drum-level indication provides a contrary indication of the water demand.

Following a sudden increase in steam demand, which causes the pressure to drop (and therefore the drum level to rise), a simple level controller would respond by reducing the flow of feed water. Equally, a sudden decrease in steam flow, which would be accompanied by a rise in pressure and an attendant fall in the drum level, would cause a level controller to increase the flow of water. Both actions are, of course, in the incorrect sense.

The effects of swell and shrinkage, in addition to being determined by the rate of change of pressure, also depend on the relative size of the drum and the pressure at which it operates. If the volume of the drum is large in relation to the volume of the whole system the effect will be smaller than otherwise. If the system pressure is low the effect will be larger than with a boiler operating at a higher pressure, since the effect of a given pressure change on the density of the water will be greater in the low-pressure boiler than it would if the same pressure change were to occur in a boiler operating at a higher pressure.

Faced with this situation, designers of control systems have responded by implementing a variety of solutions. The simplest of these is a 'two-element' system, since it is based on the use of two process measurements in place of the single drum-level measurement used above.

6.2.1 Two-element feed-water control

Remembering that the basic requirement of a feed-water control system is to maintain a constant quantity of water in the boiler, it is apparent that one way of addressing the problem would be to maintain the flow of water *into* the system at a value which matches the flow of steam *out* of it. One version of this system is shown in Figure 6.1. Here, the flow is controlled by an easily recognised device, a valve. We shall look at valves in more depth later, but for the moment assume that the version used in the diagram maintains the rate of water flowing through the valve at a figure which is directly proportional to the demand signal from the controller (i.e. if the demand signal varies linearly from 0 to 100%, the flow rate also changes linearly between 0 and 100%). Such a valve is said to have a 'linear characteristic' and in the system shown this is employed in conjunction with a transmitter that produces a signal proportional to steam flow. Used together, these two devices keep the parameters in step. If the transmitter produces a signal which is equal to the steam flow at all loads and if the flow through the valve is matched with this signal at every point in the flow range, a controller gain of unity will ensure that, throughout the dynamic range of the system, the flow of water will always be equal to the flow of steam.

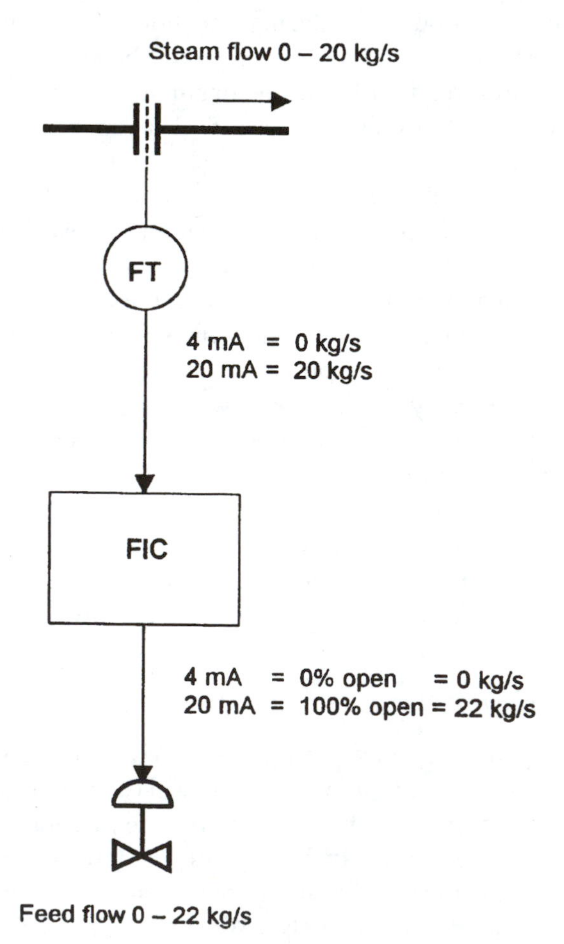

Figure 6.1 Feed-water control based on feed-flow measurement only

Naturally, scaling factors of the transmitter and the valve must be taken into account. If the range of the flow transmitter is different from the valve's flow-control range, the controller gain will need to be adjusted accordingly, and in practical systems this is always necessary.

In order to provide an adequate operational margin of confidence, the range of the control valve is always designed to be greater than the flow range of the boiler. For example, in a boiler producing 20 kg/s of steam, the valve may be sized to deliver 22 kg/s of water when it is fully open. In this example, with a linear valve characteristic, an opening of approximately 91% will be needed to pass a flow of 20 kg/s.

In this case, if the steam-flow transmitter produces an output of 100% at 20 kg/s flow, the controller gain must be such that a measured value of 100% produces an output of 91%. This is a proportional band of 110 (i.e. a gain of 20/22) and if this gain is assigned to the controller the feed flow will match the steam flow over the entire range of boiler load (assuming that the valve characteristic is linear, that the flow transmitter output is 4 mA at zero flow, and that zero flow of water occurs with a valve signal of 0%).

The problem with this system is that it only matches the steam- and feed-flow rates. If, at the outset, the drum level is below the desired value, that is where it will stay, because if everything is set up correctly the feed into the boiler will always match the steam flowing out of it, and there is no mechanism for introducing the small surfeit of feed over steam, or the slight deficit, that is needed to correct the drum-level error.

It is important to consider the practical reality of what would happen if things were *not* to be set up correctly. In this situation, if there is a small setting error in the controller gain, or if the feed valve passes more or less water than it should at the given opening, or if the steam flow transmitter is slightly out of calibration, the drum level will integrate up or down at a rate determined by the scale of the error, and nothing will correct for this undesirable state of affairs.

In other words, the system cannot correct the drum level if this parameter deviates from the desired value either because of an initial error or because of small errors in the steam-flow measurement or nonlinearities between the valve demand and the actual flow through it. In the example given above, the exact gain required is 0.909 09.... Therefore, if the controller gain were to be set to 0.91 as suggested above, the feed-water flow would be slightly greater than the steam flow, and the drum level will gradually increase.

To counter these effects it is necessary to add a feedback element, consisting of another controller which will act to correct for any mismatch

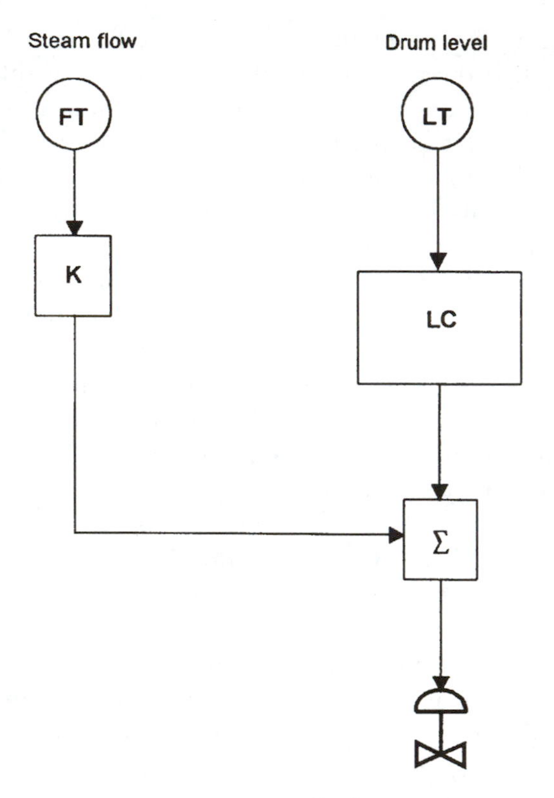

Figure 6.2 Basic two-element feed-water control system

between the actual and desired drum levels. Figure 6.2 shows one variety of such a 'two-element' system.

In such a system, because the drum volume and the steam and feed flows form an integrating system, with the drum level integrating any steam-flow/feed-flow mismatch, it is unnecessary to employ an additional integration function in the controller. Therefore the drum-level controller should be of the proportional-only variety.

The correct gain for this controller can be determined from a knowledge of the swell and shrinkage effects within the boiler. If these are not known they can be determined by test. A suitable test would be to change the steam flow as rapidly as possible by, say, 10% of the maximum evaporation rate of the boiler, while keeping the feed flow in step with the steam flow. (This can be achieved by hooking a feed-flow signal into the system while temporarily disabling the drum-level controller).

To see how the information on the boiler's swell characteristic can be used to help with controller tuning, let us examine a two-element system where the range of the steam-flow transmitter is ranged as above (0–20 kg/s), and the feed valve is again sized to deliver 22 kg/s when it is 100% open. Assume that the drum-level transmitter is ranged to produce 4 mA when the water level has dropped below the setpoint by 250 mm, and that it is 20 mA when the level is 250 mm above the setpoint (i.e. a range of 500 mm). Finally, assume that a test as described above has determined that the swell resulting from a sudden 10% change of steam flow raises the drum level by 80 mm.

If the drum-level controller is to exactly counteract the effect of swell, it must produce an output that cancels out the step change in the steam flow, which was 10%. The controller output must therefore change by 10% when the input error changes by 16%, which means that the gain must be 0.625 (10 ÷ 16).

When the steam-flow and pressure changes have settled out and the water level has returned to the setpoint, the level-controller output will again become zero. The valve opening will then revert to tracking slow changes in the steam flow, as described earlier.

This analysis depends on the swell effect being constant over the boiler load range, which may or may not be true, but it provides a practical method of tuning this type of system, and will produce a fairly good performance over a wide range of conditions. Theoretically, better results could be obtained by carrying out tests to determine the swell effect at various points in the load range and introducing a nonlinear function within the level controller to compensate for the differences across the range. But this is rather complicated for what is essentially a simple system and in any case performance is likely to be limited by the other serious deficiencies within the system, which we shall examine in the next section, which discusses a more comprehensive system, known as three-element control.

6.2.2 Three-element feed-water control

Throughout the above analysis, reference has been made to the feed-water valve characteristic being linear and the valve being sized to produce a fixed flow when it is 100% open. However, the flow through a valve depends both on its opening and on the pressure drop across it. In a feed-water system, the pressure drop across the valve varies from instant to instant, and the flow through it at any given opening will therefore vary. For reasons given earlier, in a simple two-element system based on drum level, the inclusion of an integration element in the level controller is undesirable. Therefore the varying flow results in the level control becoming offset, to restore the steam-flow/feed-flow balance. This offset is undesirable, since it needlessly erodes the safety margin provided by the presence of the drum.

One method of correcting for the error produced by the feed valve is the addition of a third element to the system—a measurement of feed-water flow.

There are various ways of implementing such a system, one of which is shown in Figure 6.3. Here, the output of the drum-level controller is trimmed by a signal representing the difference between the feed-flow and steam-flow signals. A gain block (4) is introduced to compensate for any difference between the ranges of the two transmitters. In most cases the steam-flow and feed-flow signals will cancel out, and the drum-level controller will be modulating the feed flow to keep the level at the setpoint. In this case, it is reasonable to apply an integral term in this controller, as shown.

In another implementation of this familiar system shown in Figure 6.4, a 'cascade control' technique is applied. The drum-level controller (item 5,

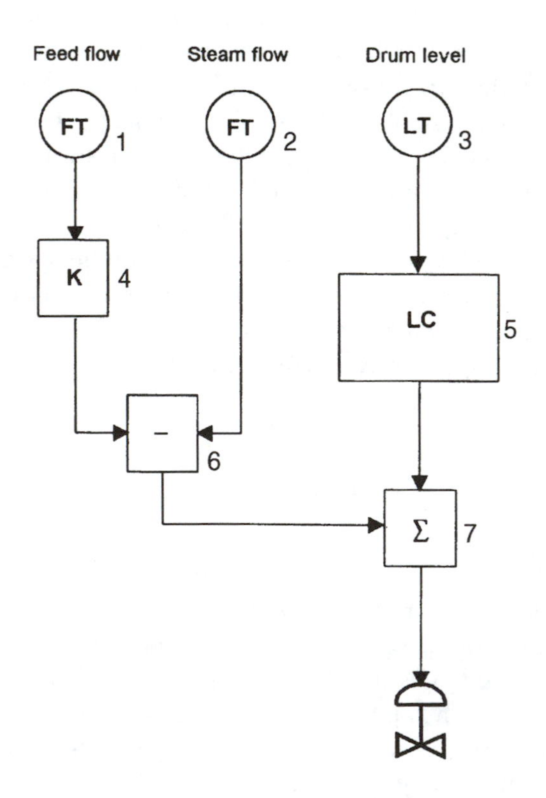

Figure 6.3 One type of three-element feed-water control system

proportional only), compares the measured level signal with a set value and produces a bipolar output proportional to any error. This trims a modified steam-flow signal, which is acting as the desired value for a closed-loop feed-water controller (7). As previously, a gain block (4) adjusts for any range difference between the steam-flow and feed-flow transmitters.

These are not the only ways of implementing three-element control. Several variants of the system are in common use, each with its own advantages and disadvantages. However, each system has one factor in common, the use of steam-flow, feed-flow and drum-level measurements. The application of the feed-flow measuring element compensates for any variations in feed-flow, whether these are due to the pump characteristics or other factors, and the three-element system is therefore recommended wherever accuracy of control is required.

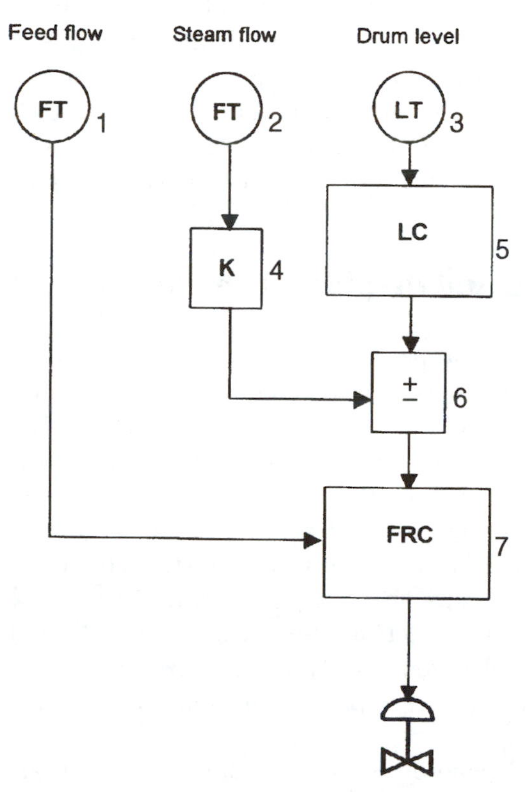

Figure 6.4 Alternative (cascade) three-element feed-water control system

As might be expected, a three-element system is more expensive than a single-element or two-element system, but it is always worthwhile to examine the real cost differences before deciding on which system to use. For example, the use of a steam-flow signal may not, in fact, add any additional cost, because this measurement may be needed elsewhere, e.g. in the combustion control system.

However, the presence of a feed-flow measurement in the system does necessarily add a significant cost burden. The transmitter itself is a complex and not inexpensive item and it may also require the provision of a primary element (such as a Venturi, flow nozzle or orifice plate) which will add further cost and require the provision of adequate lengths of straight pipework upstream and downstream of the device.

The capital cost of the system needs to be carefully considered against the background of a detailed knowledge of the operational régime of the plant. All too often, insufficient rational thought is applied to these factors. The use of a small drum in the boiler offers a not insignificant reduction in the cost of a plant, but its use also reduces the operational safety margins and therefore requires increased accuracy of feed-water control. Yet, the cost-cutting exercises that lead to the use of a small drum are very often extended to the control system as well, further reducing the margin for error. If safety margins are not to be dangerously eroded, careful consideration must be applied to the determination of practical design rules for feed-water control systems. The control implications should be given just as much consideration as the process or mechanical aspects.

6.3 Measuring and displaying the drum level

From the above, it is apparent that the primary objective of the feed-water control system of a drum-type boiler or HRSG is to maintain the drum level at the correct value. We shall now look at how this parameter is measured. It is an area where the problems are unexpectedly complex.

Figure 6.5 shows one method of measuring the drum level. This connects the differential-pressure transmitter directly to the drum, via isolating valves. Note the 'constant-head' reservoir connected to the upper tapping point. Because the impulse pipework to the transmitter is outside the heated zone of the boiler, any steam within it will tend to condense, and the pressure applied to the HP port of the transmitter will therefore be the steam and water pressure plus the pressure due to the weight of this condensate.

The latter will depend on the volume of condensate that has collected, and this will be time-dependent. The pipe will be full of steam after the

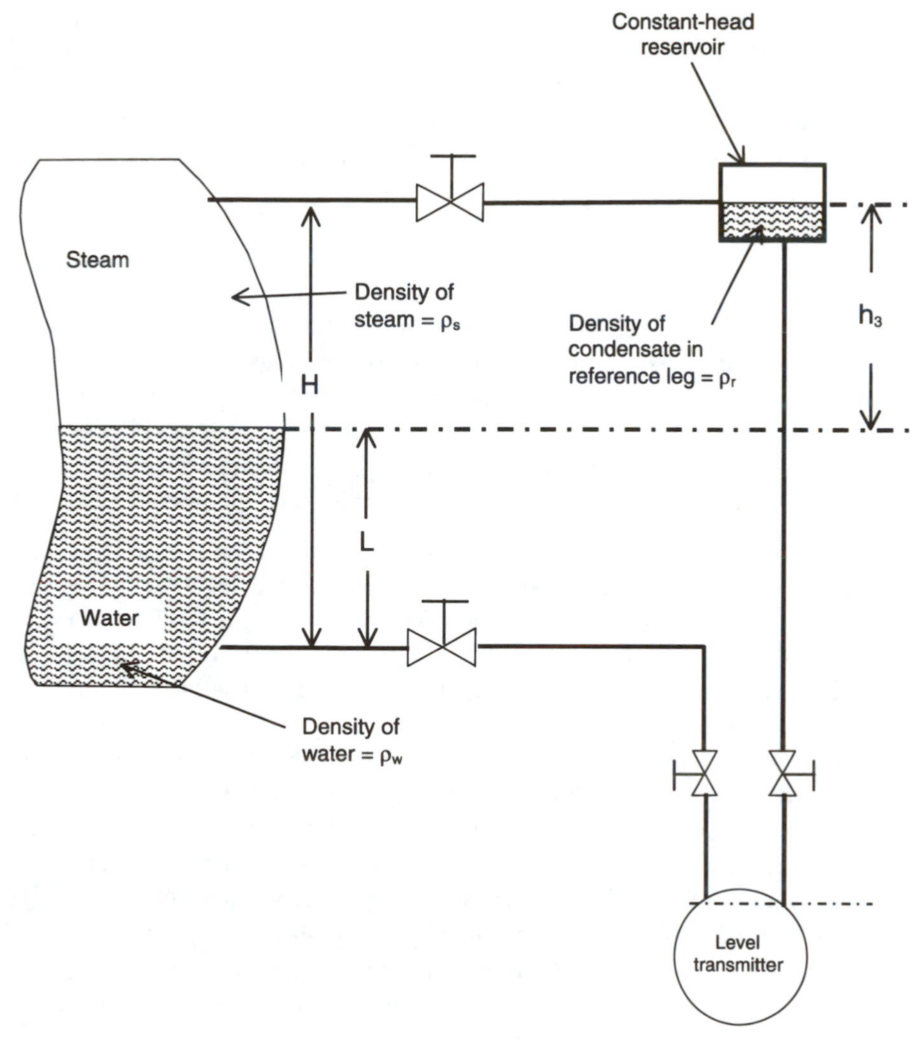

Figure 6.5 Measuring the drum level without a water column

transmitter impulse lines have been 'blown down' on start-up, but it will afterwards start to fill with condensate. The constant-head vessel is left deliberately unlagged, so that the steam in it condenses, maintaining the pipe to the transmitter full of condensate.

The differential-pressure appearing at the transmitter ports is a function of several variables: the water level in the drum, the densities of this water and the steam above it and the density of the water in the pipes to the transmitter. In addition, the derivation of a level signal needs to take into account the density of the fluid used during the initial calibration of the transmitter. Finally, it is also necessary to recognise the requirement that the lowest drum-water level at which the system will be allowed to operate will be at some point above the level of the lower tapping point.

The differential pressure at the transmitter is defined by*

$$\mathrm{DP} = (1/\rho_r) \times [(H \times \rho_c) - [(L \times \rho_w) + (H - L) \times \rho_s)]]$$

where

DP = the differential pressure at the transmitter (at 20 °C)
H = the distance between the bottom and top tappings
L = the height of water in the drum above the bottom tapping
ρ_r = the density of water used for calibrating the transmitter (at 20 °C)
ρ_s = the density of the water in the drum at the operating pressure
ρ_w = the density of the steam in the drum at the operating pressure

It should be noted that the differential pressure at the transmitter will be highest when the separation between the water level and the top tapping is the greatest i.e. at low levels. In other words, a 4–20 mA transmitter will produce 4 mA at the highest level and 20 mA at the lowest. This apparently reversed 'sense' must be corrected by the DCS before the density calculation is applied.

Since the densities of the steam and water in the drum will both depend on the conditions that exist in the drum, the pressure needs to be taken into account when calculating the actual level based on the differential head produced at the transmitter. At one time, various proprietary devices were available for performing the required calculation, but it is most economical, these days, to perform the calculation in the DCS, based on the differential-pressure and pressure signals. In this case the above equation will enable the level to be calculated for any combination of signals from the two transmitters.

However, an important point to bear in mind when using the DCS to perform the pressure compensation in this way is that the corrected signal will be available to the operator *only while the DCS is operational.* If a major failure should occur in the computer system, it is important that the operator can still be able to monitor the drum level by other means, and these must be compensated in a similar way to the above, so that the indication is relatively unaffected by pressure.

Figure 6.6 shows a proprietary system which generates a drum-level signal for display locally and in the control room. This technology is based on the fact that the conductivity of water is different from that of steam,

* Courtesy of Bristol-Babcock Ltd, Kidderminster, Worcestershire, UK

(a)

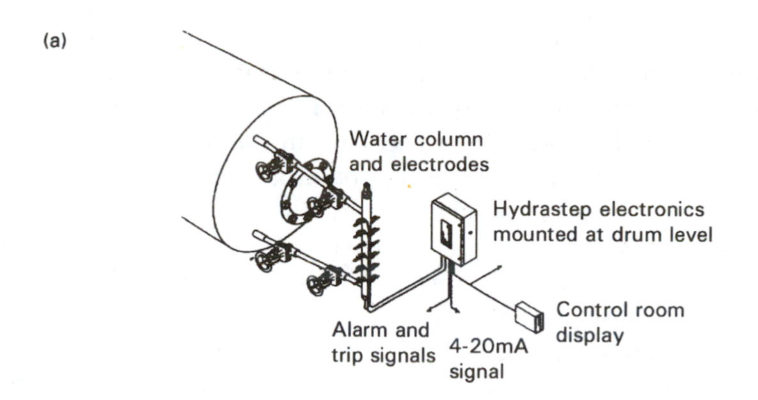

(b)

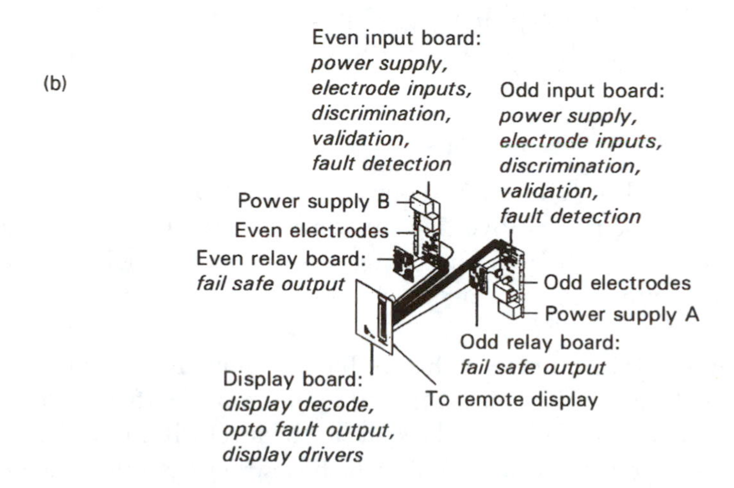

(c)

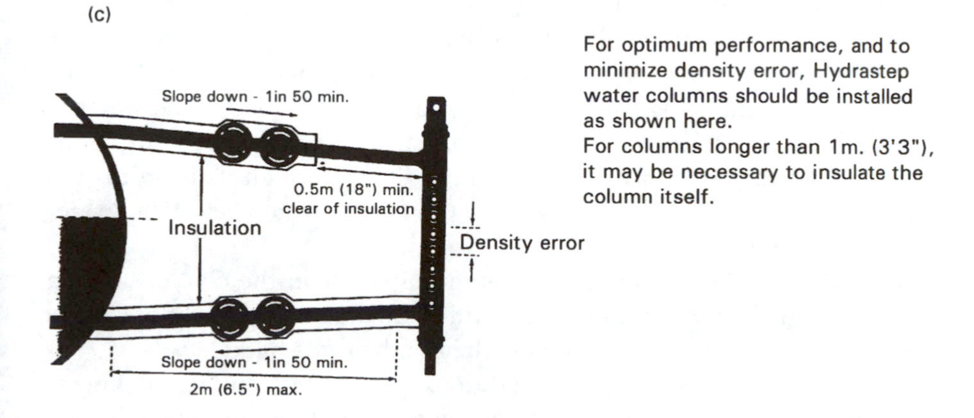

For optimum performance, and to minimize density error, Hydrastep water columns should be installed as shown here.
For columns longer than 1m. (3'3"), it may be necessary to insulate the column itself.

Figure 6.6 The 'Hydrastep' level-indicating system

and a series of electrodes mounted in a water column attached to the drum uses this fact to detect the interface between water and steam. (The use of an external water column is necessary because of the number of penetrations that would otherwise be necessary in the boiler drum.) The detection circuit is divided into two groups of 'even' and 'odd' electrodes, so that failure of a single drive circuit cannot disable the entire system.

Although such a system will be affected by the difference in the densities of the water in the drum and the column, careful design of the installation will minimise any errors (Figure 6.6*c*).

At present, although such devices provide an excellent indication of the drum level, they are not suitable for control, because of the transient disturbance that occurs as the level moves from the position of one electrode to another. These step changes can produce unpredictable effects in the control loop.

6.3.1 Using an external water column

Although the method of connection shown in Figure 6.5 is viable, it has the disadvantage of being sensitive to errors during sudden reductions in the boiler pressure, caused by the condensate 'flashing off'—boiling as the temperature of the fluid suddenly finds itself above the saturation temperature.

An arrangement that minimises this problem is shown in Figure 6.7, where an external water column is connected to the drum, so that the level of water in column is (theoretically) the same as the level within the drum. The column uses a volume of stored fluid which is larger than the volume of condensate in the small-bore HP leg of Figure 6.5, and the system is therefore less vulnerable to flashing off.

As with the detection column shown in Figure 6.6, great care must be taken to avoid errors caused by the temperature of the fluid in the measuring system being very different from that of the steam and water in the drum. This leads to a density error, since the water column at the gauge will balance with the level in the drum, although its length is less than the distance between the lower tapping and the level of the water within the drum.

Considering the drum and the column of water in the column to be a 'U' tube, it will be seen that balance will occur when the weight of fluid in the left-hand leg equals the weight of fluid in the right-hand leg.

By keeping the temperature conditions within the column as close as possible to those in the drum, the density error will be minimised. This is done by arranging the pipework so that fluid flows through the water column. Unlike the drum itself and all the other pipework, the water

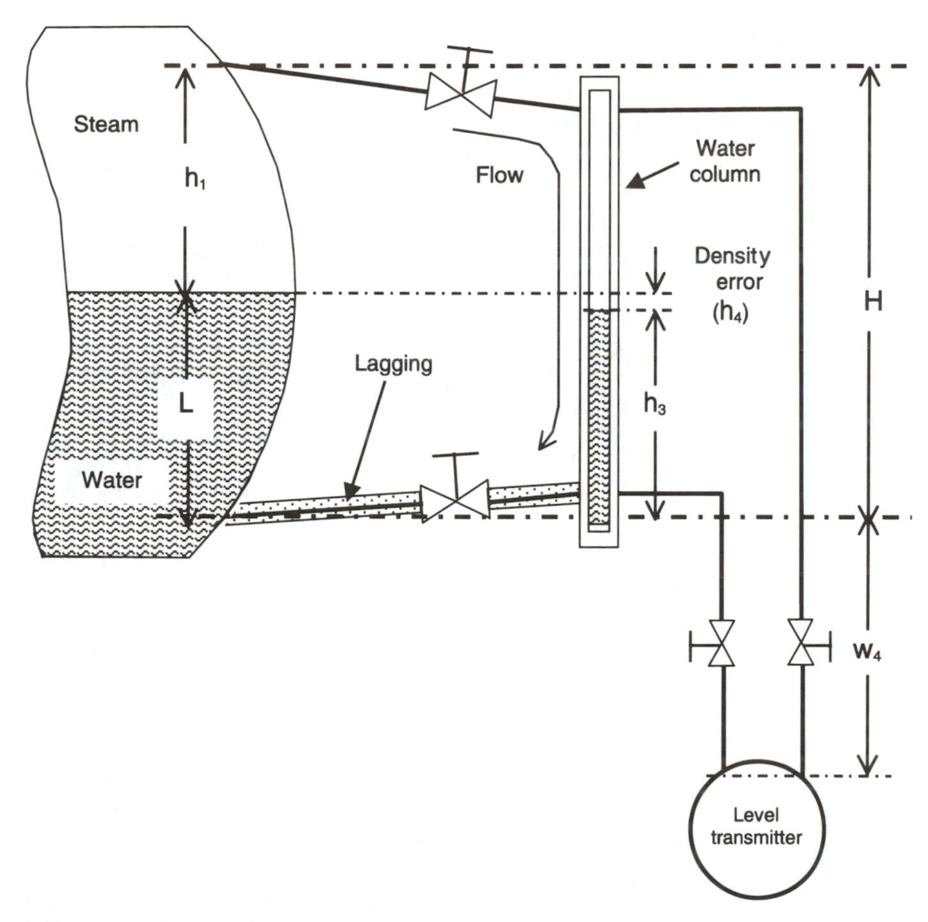

Figure 6.7 Drum-level measurement with a column

column and the section of pipe connecting it to the tapping-point isolation valve (on the right in Figure 6.6) are left unlagged so that the steam condenses and the impulse line remains full of condensate. A circulation of fluid is then established through the column, with steam leaving the drum at the upper tapping point, condensing in the pipe and water column, and returning to the drum via the lower tapping point. This flow tends to maintain the temperature within the water column at a value which is as close as possible to the condition within the drum. Nevertheless, some temperature difference will still exist, and this will have the effect of increasing the density of the fluid in the cooler parts of the system.

6.3.2 Statutory requirements

In many countries, there is a legal requirement to provide separate systems to make the boiler operator aware of the level of the water in the drum. In the UK, British Standard BS 1113 : 1998* refers to this subject in section 7.3, as follows:

> Each steam boiler, in which a low level, or into which a low flow rate, of water could lead to unsafe conditions shall have at least two independent and suitable means of indicating the water level or flow. Each indicating device shall be capable of being isolated from the boiler and each device shall be a water level gauge in which the water level can be observed except in the following cases:
>
> (a) ... (refers to once-through boilers)
> (b) For boilers with any safety valve set at or above 60 bar *g*, the use of two independent manometric remote water level indications shall be permitted in place of one of the water level gauges. In such cases these remote water level indicators shall have their own independent connections to the boiler.
> (c) For boilers of less than 145 kg/h evaporative capacity, one water level gauge is sufficient.
> (d) The use of alternative devices in place of water level gauges in which the water level can be observed shall be permitted, subject to agreement between the manufacturer and the Inspecting Authority. The design of the devices shall combine appropriate design principles such as fail-safe modes, redundancy, diversity and self-diagnosis in order to provide suitable and reliable indication.
>
> The water level gauge in which the water level can be observed shall be mounted so that the lowest water level that can be observed is at least 50 mm above the lowest water level at which there will be no danger of overheating any part of the boiler, when in operation at that level. Where this is not practicable, the water level gauges shall be sited by agreement with the Inspecting Authority in positions that have been found by experience to indicate satisfactorily that the water content is sufficient for safety under all service conditions.
>
> At least one water level gauge with its isolating valves or cocks shall be connected directly to the boiler, and other than a drain, no device shall be fitted to the gauge that could cause incorrect indication of the water level in the gauge.

Figure 6.8 is an attempt to illustrate some of the above requirements pictorially. Note that the aim of this diagram is only to illustrate the requirements and to provide general guidance. It is important that a

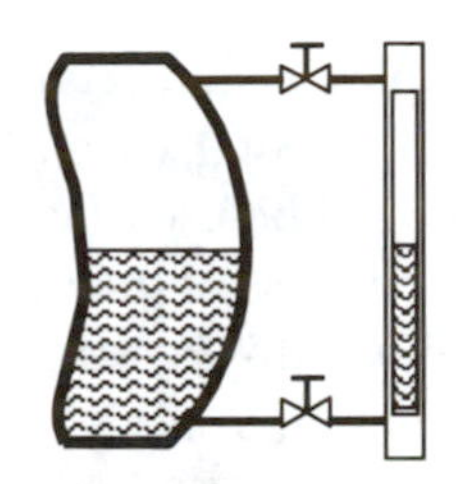

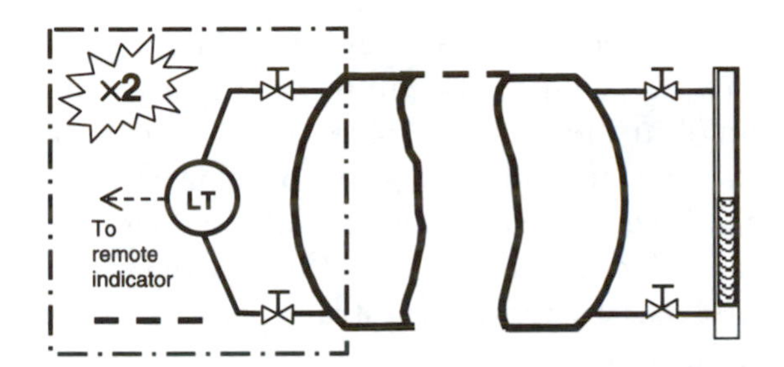

A 'water level gauge in which the level can be observed ...'.

Note that at least two such gauges are required on boilers of 145 kg/h (0.04 kg/s) or above.

'For boilers with any safety valve set at or above 60 bar *g*, the use of two independent manometric remote water level indications shall be permitted in place of one of the water level gauges. In such cases these remote water level indicators shall have their own independent connections to the boiler.'

'....The use of alternative devices in place of water level gauges in which the water level can be observed shall be permitted, subject to agreement between the manufacturer and the Inspecting Authority. The design of the devices shall combine appropriate design principles such as fail-safe modes, redundancy, diversity and self-diagnosis in order to provide suitable and reliable indication.'

Figure 6.8 Permitted alternatives for water-level measurement on drum-type boilers and HRSGs
Based on BS 1113 : 1998

specific installation is designed to meet the standards required at the actual point of use, since the standards set by authorities in the relevant country, or by the insurers, may differ substantially from those indicated by this diagram.

The above Standard also refers to low-level alarms, stating that

> Every steam boiler shall be provided with a low water level or low water flow audible alarm device.

> Water level alarms, whether of low water or high and low water type, shall be so fitted that the alarm is actuated while the level is still visible or indicated in the water level gauges.

With reference to the size of the connecting pipework BS 1113 : 1998 states that:

> Where a water level gauge, safety control or alarm device is connected to the boiler by pipes, the bore of such pipes shall not be less than 25 mm.

(The standard allows the ends of the pipes local to the fittings to be reduced to not less than 20 mm bore for water level gauges and to 25 mm bore for separate safety control and alarm devices.)

In this respect the Standard differs from others, but on one point there is agreement, there is a need to use comparatively large-bore pipework for instrument connections. The use of such pipes increases the cost of the installation, but their use is important from the viewpoint of safety, since small-bore pipework can become obstructed (for example by sludge) and an obstructed impulse pipe can lead to an instrument providing seriously incorrect information on the parameter being measured. (The use of very small-bore pipes can also cause a reading to be sluggish because of the time taken for pressure changes to affect the instrument, though this is less of a problem with today's instruments which displace very small volumes for full-scale operation.)

6.3.3 Discrepancies between drum-level indications

It sometimes happens that various instruments connected to the same boiler drum display level measurements that are significantly different from each other. Since it is unlikely that the actual drum is anything but horizontal (except for installations on submarines during diving operations) such discrepancies must be due to some error or other. The following list summarises the factors that can cause errors:

- *Density errors*: Differences between installations can cause one instrument to be more affected by density factors than another. (One possibility is that, inadvertently, lagging has been applied to one of the condensation reservoirs).
- *Turbulence*: The surface of the boiling water inside the drum is anything but still. It has been known for 'standing waves' to exist around the downcomers, affecting some measurement points more than others.
- *Flashing-off*: Differences in the geometry of the measuring systems can cause some measurements to be more affected than others by flashing-off during pressure changes.
- *Calibration*: It is vital that all transmitters are carefully and accurately calibrated, and that any density compensation is correctly set up.
- *Installation*: As stated earlier, errors or sluggish response can be the result of partial or complete plugging of impulse lines, or imperfect blow-down operations.

6.3.4 Steam extraction

In both the two-element and three-element systems, an assumption made in the above examples is that the steam output by the boiler eventually returns to the inlet in the form of water. This is not true where significant losses occur or if any steam is abstracted for applications such as soot-blowing. Here the steam is effectively lost, and if a soot-blower abstraction is made at the drum the amount being used will not be included in the flow measurement. This will result in the drum level being offset from the desired value since the flow of feed water into the system should be equal to the steam taken by the load (plant or turbine) plus the steam used for soot-blowing. The effects of such operational factors will be particularly significant where the design of the control system has been based on the use of a proportional-only controller. In such cases it may be necessary to add a small degree of integral action to the controller, although this should be restrained since the additional integral action can affect the stability of the system.

6.4 The mechanisms used for feed-water control

In this analysis we have looked at the principles of control, and seen that because of various problems, a variety of control methods has evolved, the selection of which depends on a variety of engineering and economic considerations relating to each application. In the discussion some reference had to be made to the mechanism for controlling the flow and for simplicity it was assumed that this was by means of a familiar device, a valve. Now we shall look at the nature of valves in greater depth, and then we shall examine other methods of controlling the flow.

6.4.1 Valves

What follows is merely a practical overview of valve designs in general. It is not intended to be a deep analysis of what is in itself a specialised subject. If more detailed information is needed, it can be obtained from the many textbooks on valve design or from the publications produced by various valve manufacturers. (See further reading section of this book for information on three such books.)

A control valve consists of many components which may conveniently be considered as falling into one of two groups: the valve body and the actuator. The former is the part through which the water flows and this flow is controlled by adjusting the resistance offered to the water. This is done by moving the position of a plug in relation to its seat. The position of the plug is controlled by an actuator which acts via the stem.

Figure 6.9 shows a small-bore feed-water control valve body with a contoured trim (the 'trim' being the part of the valve which is in flowing contact with the water). The contour determines the relationship between the position of the plug and the flow of water past it. The type of trim will be dictated by the application, such as the need to minimise acoustic noise or cavitation, the rangeability needed etc. In addition the trim design will determine the valve *characteristic*, which is the curve relating the stem position to the rate of flow of water through the valve. This is an important feature, since the characteristic determines the gain of the valve system, which forms part of the overall loop gain.

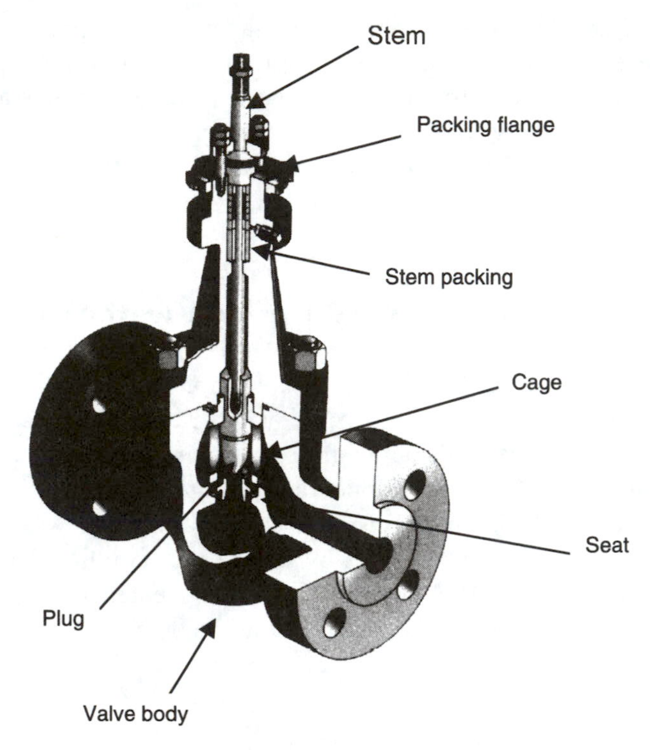

Figure 6.9 A typical feed-water control valve body
© Hopkinsons Ltd. Reproduced by permission

For a given opening, the flow through the valve will be determined by the delivery pressure of the feed pump and the resistance that the boiler pipework offers to the flow. To simplify the task of selecting the correct valve size and characteristic, it is necessary to relate everything to a definable set of conditions. This is achieved by determining what the flow through the valve would be if a fixed differential pressure were to be maintained across it. This is termed the *inherent characteristic* of the valve.

Once the valve is operating on the actual plant, the position/flow relationship achieved in practice will not match the inherent characteristic, because in the real world the inlet pressure and system resistance will vary, producing a pressure drop which is different from the value that was used to define the inherent characteristic. The pressure/flow relationship achieved in actual operation is called the *installed characteristic.*

As stated earlier, the gain of the valve is initially defined by the inherent characteristic, three types of which are commonly available, as shown in Figure 6.10. The operation of these different characteristics is now examined.

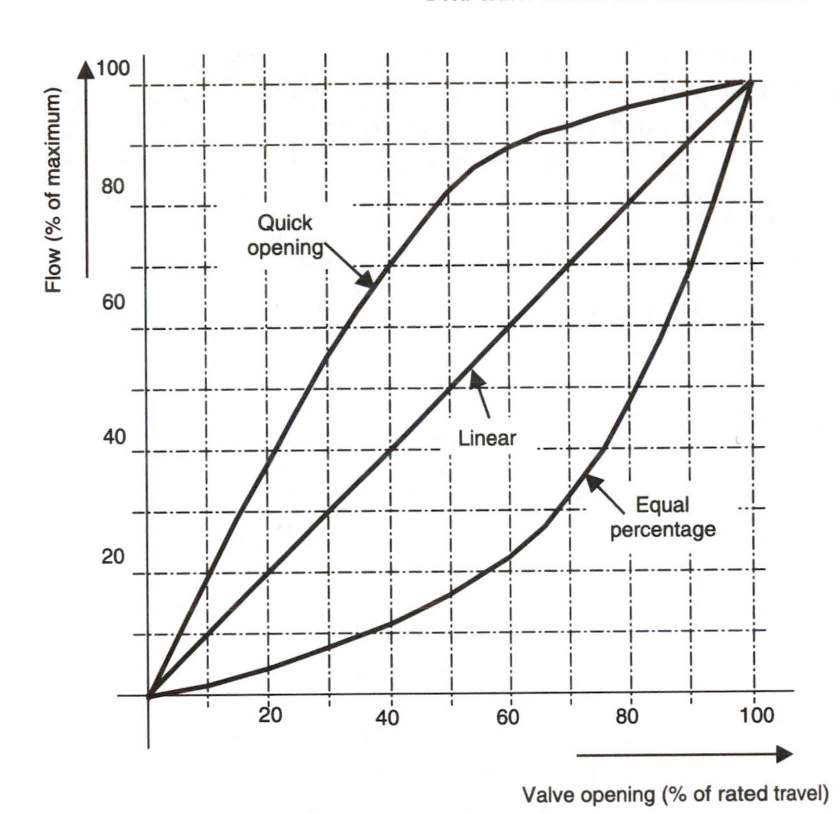

Figure 6.10 Inherent characteristics of valves

6.4.2 Quick-opening

With a quick-opening valve, the flow rate through the valve changes very rapidly at low openings, with a slope that is fairly linear. Once the valve has opened about half way the rate of change of flow diminishes. This type of characteristic is usually applied only to shut-off valves.

6.4.3 Linear

When a valve has the linear characteristic, the flow rate through it at any given opening (in terms of percentage of maximum flow) is directly equal to the valve stem position (as a percentage of its full travel). With this type of characteristic, the gain of the valve system is constant for all openings. However, the flow through the valve at any given opening depends on the pressure-drop across it and the linear characteristic applies *only if the pressure drop across the valve is constant for each opening*, a condition that may not exist in practice unless special attention is paid to achieving it.

As a general rule, a linear characteristic is preferred for feed-water control applications, since it simplifies tuning of the loop and enables good performance to be achieved over the widest possible range of flows.

Where optimum performance and efficiency is required the additional cost of providing a constant pressure drop across the valve should be considered. Such a solution will add cost, but this will be offset by improved control performance and plant life and by the savings achieved by not running the feed pump or pumps at a higher speed than necessary at reduced loads.

6.4.4 Equal percentage

The third characteristic is called equal percentage. Here, for all positions of the stem, the flow change achieved by moving the stem by a given amount is a constant proportion of the previous flow. What this means is that a given stem movement will change the flow by the same ratio *of the previous flow*, at any point in the valve travel. Therefore, the larger the opening, the greater will be the change of flow produced by a given stem movement.

This is shown by the curve in Figure 6.10, where moving the stem from 20% of full travel to 30% of full travel changes the flow from 5% to 7.5%, while moving the stem from 80% of full travel to 90% of full travel increases the flow from 50% to 75%. In both cases, a stem movement of 10% of full travel results in a 50% increase of the previous flow.

From this it is apparent that, with this type of characteristic, the change of flow for a given stem movement is smallest at low openings and greatest at high openings. In other words, the gain is lower at low flows than it is at high flows. Equal-percentage valves are used where the mechanical plant design is such that there is only a small pressure drop available for the valve, or where the pressure drop across the valve is likely to fluctuate over a wide range.

6.4.5 The valve sizing coefficient

Feed valves are designed to pass a flow that corresponds to the maximum flow requirement of the boiler plus a safety margin, and the capacity of the valve is related to a factor known as the valve sizing coefficient. The factor used widely across the world is based on US units and relates the capacity of the valve to the pressure drop by the following formula:

$$Q = Cv\sqrt{(\Delta P/G)}$$

where

Cv = sizing coefficient
Q = capacity of valve in US gallons per minute (1 US gallon = 3.785 litres)
ΔP = pressure differential across valve in pounds per square inch (psi). (1 psi = 0.069 bar)
G = specific gravity of the flowing fluid

The sizing coefficient is determined by experiment for each type and size of valve using water as the test fluid. It is equal to the volume of water (in US gallons) that will flow through the wide-open valve in one minute when the pressure drop across the valve is 1 psi. The European equivalent of the sizing coefficient is known as Kv. To convert between these units multiply the Kv figure by 1.66 to obtain the Cv.

Given the pressure, temperature, flow and line-size characteristics for a given feed-water application, valve manufacturers will be able to provide detailed guidance on the correct valve size for a given application and from this will be able to predict the pressure drop across the valve.

6.4.6 Fail-safe operation

In the course of designing a feed-water control system, another matter that must be considered is that of selecting the 'fail-safe' position of the plug, the state that will arise if the actuator fails, or if the command signal to the valve is lost. The actual selection will depend on a range of factors, but in determining the safest option it is important to consider the effects of the flow on the valve itself, since the applied pressure may tend to force the plug open or closed. Once again, valve manufacturers will be able to provide advice on this matter in relation to the actual installation being designed.

6.4.7 Selecting the valve size

The size of valve to use in a given application will be determined by many factors, only one of which is the physical size of the line in which it is fitted. Clearly, the valve must be large enough to pass the required flow with ease. Oversized valves will be unnecessarily expensive and should be avoided as unable to effectively control small flows.

For a valve to maintain any control over the process there must be some pressure drop across it, but if the pressure drop is too great a number of undesirable events start to occur.

Figure 6.11 shows in profile the pressure along a short section of pipe containing a restriction (such as a valve). Two curves are shown, one for a valve where careful design produces streamlined flow and minimal pressure loss, and another for a valve through which the flow is more,

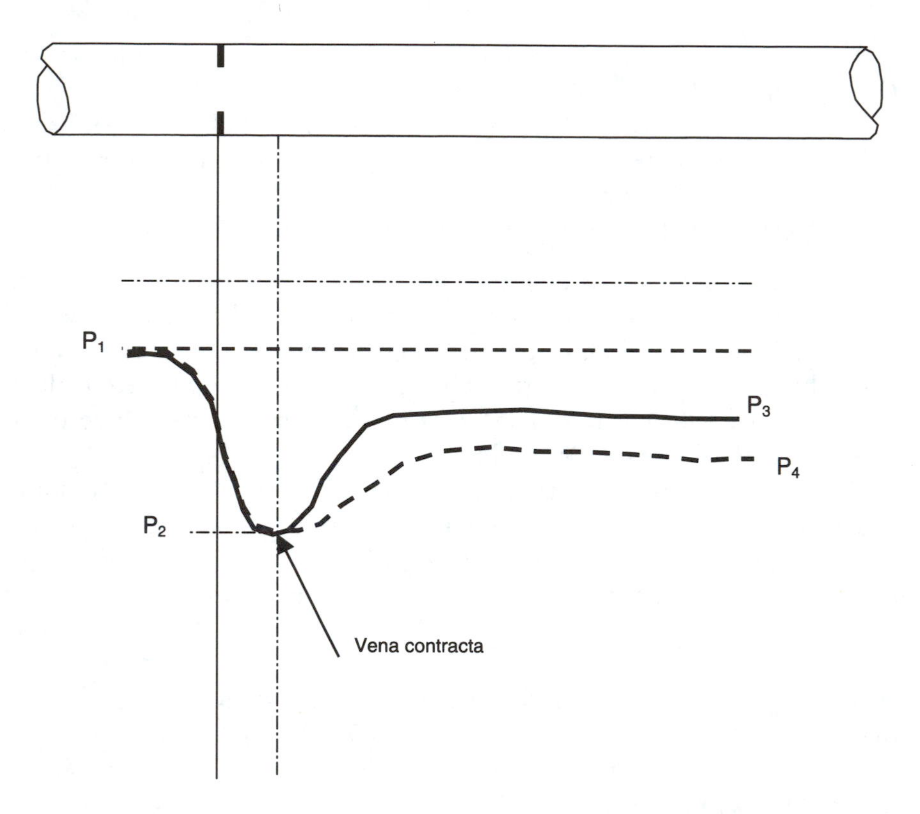

Figure 6.11 Pressure profile along a pipe containing a restriction

producing a greater pressure loss. In each case, the lowest pressure occurs a short distance downstream from the restriction (the Vena contracta) and the pressure rises past this point although it obviously never regains the initial value. If the pressure at the inlet (P_1) is such that the pressure along this curve drops below the vapour pressure, the fluid will start to boil and bubbles will form in it at the point where this occurs.

One of two things will occur as the pressure rises again after the restriction. If it rises above the vapour pressure, the bubbles will collapse—a process known as cavitation, which is accompanied by acoustic noise, the degree of which depends on the scale of the cavitation. Apart from being undesirable on environmental grounds, this noise represents an expenditure of energy. Furthermore, if the bubbles collapse close to metal surfaces, the localised energy release can damage the metal. In some cases this will become apparent as severe pitting of the valve plug and cage, or it could appear as pitting of the pipe itself if the point where the vapour pressure threshold is passed occurs some way down from the valve. High recovery valves are more likely to experience cavitation, because the downstream pressure is more likely to exceed the vapour pressure.

If, on the other hand, the pressure does not rise above the vapour pressure the bubbles will remain suspended in the fluid, to be carried downstream of the restriction. This is known as 'flashing', and it can result in erosion damage to the valve internals at the point of maximum velocity (usually at or near the point where the plug seats against the ring).

These bubbles reduce the valve's ability to pass fluid and eventually it 'chokes'. When this occurs no more flow can occur irrespective of how much water pressure is applied at the valve inlet. From these considerations it is apparent that the production of bubbles may cause noise and damage to the valve, and possibly the pipework, and it will be correctly surmised that an optimum set of conditions will exist for a given design of valve. In other words, the valve will work best at one pressure-drop point.

To summarise, if the supply of water to a boiler is controlled by throttling the flow through a valve, this can cause erosion and noise and although control can only be effected by maintaining some pressure drop across the valve, this loss represents a loss of energy which should be reduced to the minimum.

6.5 Pumps

The water flowing through the feed valve into the boiler is delivered at pressure by one or more feed pumps. These produce a head of water which is related to the flow through the pump by a characteristic that will be similar to Figure 6.12, which shows that although the discharge pressure remains relatively constant as the flow rises from zero to (in this example) about 50%, beyond this value the pressure tends to decay as the flow increases. From this it will be apparent that the feed valve has to produce a greater pressure drop at low loads than at high loads.

Figure 6.13 shows the flow/pressure characteristic of a pump delivering water into a fixed system. In practice, as the flow through the system increases the resistance offered to it also increases, as shown by the dotted line in the diagram.

At any flow, in order to deliver water into the boiler, the pressure drop across the control valve will be the difference between the pump delivery pressure and the system resistance. For much of the flow range the pressure drop will inevitably be greater or less than the optimum for the valve design. This can be overcome by changing the inlet pressure so that the

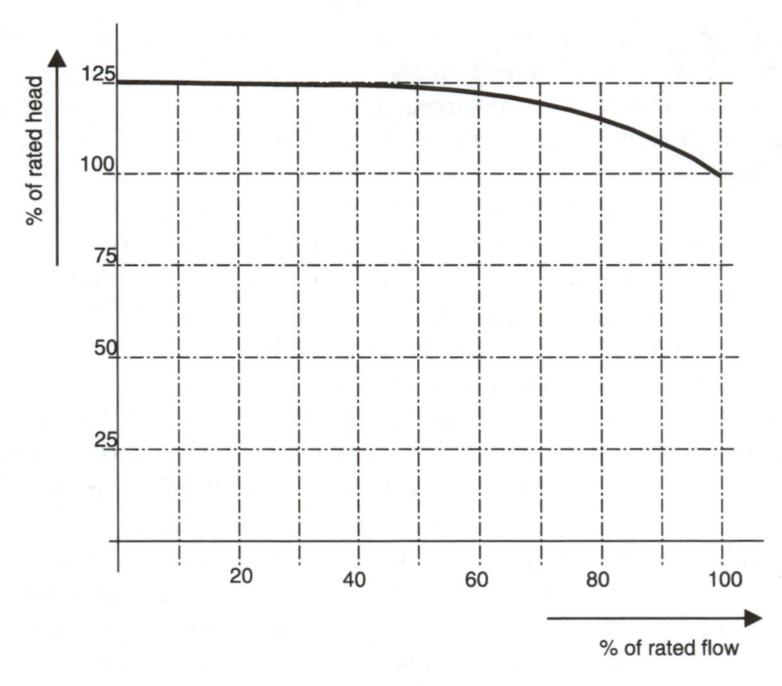

Figure 6.12 Typical speed curve for a feed pump

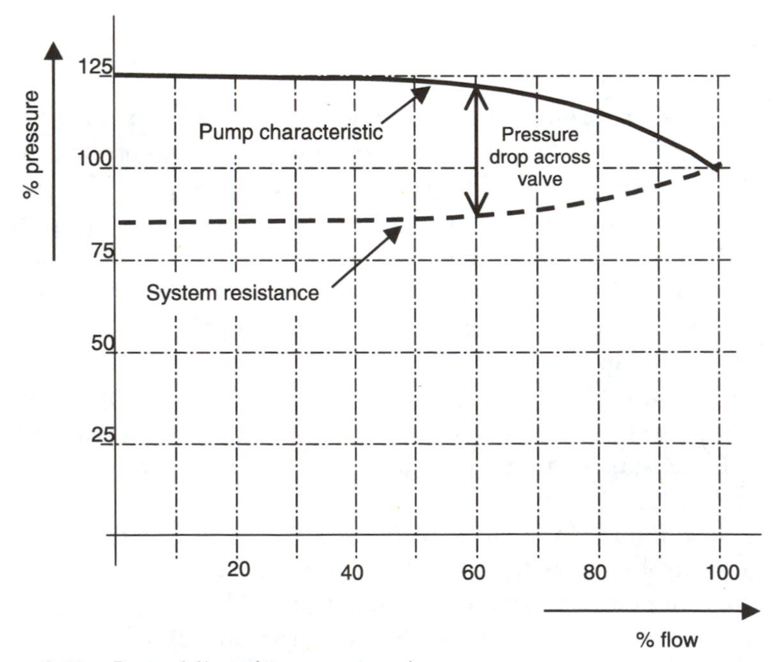

Figure 6.13 Pump delivery into a system resistance

pressure drop always remains at the optimum value. This is achieved by the use of variable-speed pumps.

6.5.1 Variable-speed pumps

Although in many cases the feed pumps operate at a fixed speed, at the design stage consideration should be given to the option of using variable-speed pumps (sometimes known as 'controllable-speed' pumps) because these will enable the feed valve to operate at the optimum pressure conditions for all loads.

The characteristics of a variable-speed pump are shown in Figure 6.14. The four curves show the pressure/flow characteristics at four different operating speeds (A–D), with A being the slowest and D the fastest speed. From this set of curves it will be appreciated that by adjusting the speed of the pump it will be possible to maintain a fixed differential pressure across the feed regulating valve at all boiler loads. In the example shown the pressure drop across the valve (δP) achieved by operating the pump at speed A at 70% flow is maintained at the same value by operating the pump at speed B at 80% flow, speed C at 90% flow etc., as indicated by the lines with arrows at each end.

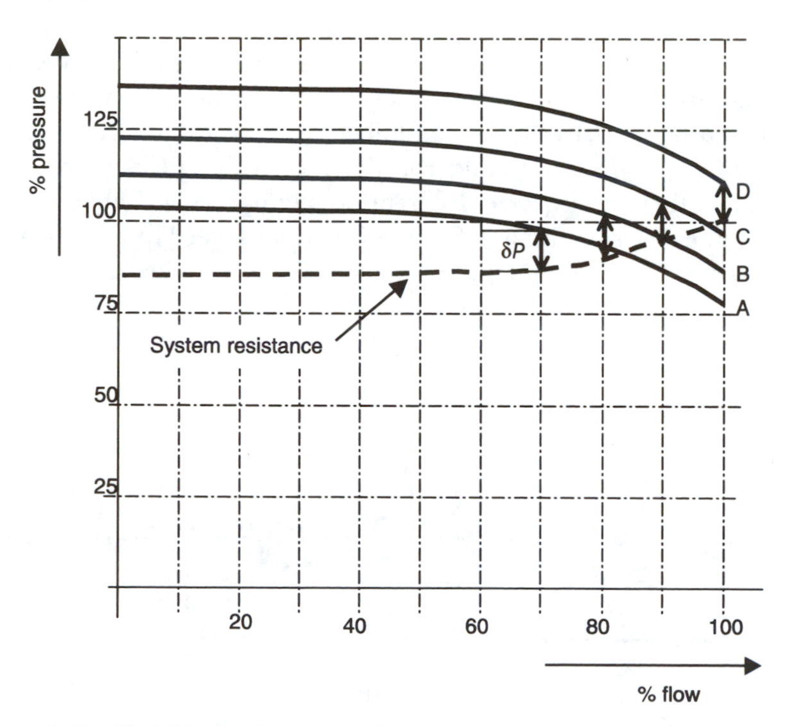

Figure 6.14 Variable-speed pump operation

The advantages of using a variable-speed pump include:

- Improvement of efficiency because of reduced pressure loss.
- Reduction of pumping power.
- Reduction of feed-valve wear due to erosion when operating at low flows.
- Improved control because the valve operates at its designed pressure-drop.
- Improved control offered by the ability to operate with constant loop gain.

Variable-speed pumps are more expensive than fixed-speed ones, but the increase in capital cost is offset by the revenue savings that will be gained, particularly if the boiler operates at reduced throughputs for a significant time over its life. Careful financial analysis will determine whether the savings do justify the additional capital cost, but these calculations must be based on assumptions about the operational regime, and these may change under the influence of external factors which are difficult to predict at the design stage.

Of course, the pumps, valves and boiler pipework are only part of the overall system. Where spray attemperators are used (see Chapter 7), the feed pumps must also be capable of delivering cool water to the nozzles. The design of attemperators requires the water to be at delivered at a pressure which exceeds the steam pressure by a minimum value. Bypassing the pressure drop across the system as shown in Figure 6.15 allows sufficient differential pressure to be maintained at the spray nozzle. But, if a variable-speed pump is used and run down to a low delivery pressure, there is a risk that the required differential may not be available. In such situations, the decision to use fixed- or variable-speed pumps will be affected by the need to maintain an adequate water pressure at the attemporators.

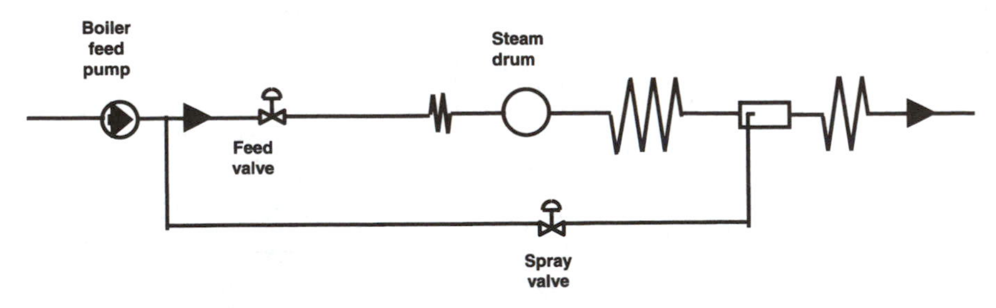

Figure 6.15 Maintaining adequate spray-water differential pressure

6.6 Deaerator control

Strictly speaking, control of the deaerator is not a function of feed-water control which is the subject of this chapter. However, as the deaerator is an essential link in the feed-water supply system it is appropriate to consider its control systems here.

In Figure 2.5 we saw how steam admitted to the deaerator rises upwards past metal trays over which the water is simultaneously cascading downwards. As the water and steam mix and become agitated, entrained gases are released. The dissolved gases are vented to the atmosphere because the vessel is pressurised by the steam. The deaerator is situated in the water circuit between the discharge of the condenser extraction pump and the inlet of the feed pumps, as shown in Figure 6.16.

It will be evident that two control functions are required by the deaerator: one to maintain the steam pressure at the optimum value, the other to keep the storage vessel roughly half full of water.

6.6.1 Steam pressure control

The pressure of the steam entering the deaerator is maintained by a simple controller whose measured-value signal is obtained from a trans-

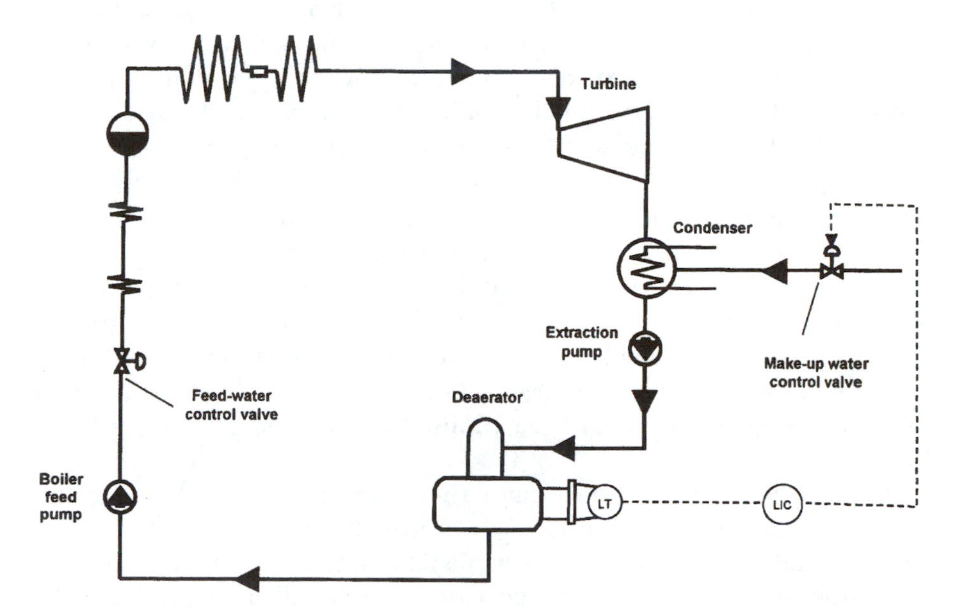

Figure 6.16 Principle of deaerator level control system

mitter measuring the steam pressure in the deaerator. The set value of the controller is normally fixed.

It has already been explained that the steam supply may be obtained either from the boiler or from an extraction point on the turbine. If the latter source is used, special consideration has to be given to ensuring that an event such as a turbine trip does not deprive the deaerator of the steam supply it needs for its operation. This purpose is served by taking a steam supply from the boiler and passing it to the deaerator via a pressure-reducing and desuperheating system (PRDS). This steam supply is referred to as 'pegging steam'. Judicious adjustment of the PRDS controller setpoint will ensure that an adequate steam supply for the deaerator is obtained at all times. However, to ensure rapid response to a turbine trip, a system of interlocks should be provided, so that the pegging steam supply is brought into service immediately on detection of a trip.

6.6.2 Level control

The storage vessel provides a measure of reserve capacity for the plant. To achieve this function the level of water in it must be maintained at roughly the midpoint. This is achieved by means of a level controller whose measured-value signal is obtained from a differential-pressure transmitter or from capacitive probes which would normally be connected to tappings of an external water column which is in turn connected to the top and bottom of the deaerator storage vessel. If there were no losses in the system, the amount of water would be constant and the level in the deaerator storage vessel would remain at the correct value set during commissioning. However, losses are inevitable (for example, due to leakages at pump glands or during soot-blowing or blowdown operations), and a supply of treated water must therefore be made available. The deaerator level controller output adjusts the opening of a valve that admits this make-up water to the condenser, as shown in Figure 6.16.

The make-up supply is conventionally fed into the system at the condenser. Figure 6.16 shows that interaction between the level controllers of the deaerator and condenser is inevitable. The situation is made more complex because the condenser extraction pump has to be provided with a bypass arrangement to maintain a minimum flow through the pump at all times.

In fact, the conditions which cause the deaerator level controller to call for more water to be added to the system will also cause the condenser level to fall, and so the two systems do not act in opposite senses. Nevertheless they do interact, and care must be taken to minimise the instability that is likely to arise.

6.6.2.1 Spill control

In addition to the level control system described above, a system must be provided to drain excess water from the deaerator storage vessel in the event of an oversupply of water and steam. This function is normally achieved on large power-station boilers by using the output of the level controller to operate the make-up and spill valves in split range. A typical arrangement is shown in Table 6.1. Smaller boilers will be limited to having a spill valve whose opening is normally controlled on an on/off basis from a high-level alarm.

6.6.3 Integrated level control

The long dead time and long time constant of the deaerator level system complicates the design and commissioning of this control loop. This situation has been worsened by the commercial pressure to operate power-station boilers at partial load and the increased need for two-shifting operations. (The term 'two-shifting' refers to operations where the plant runs for two eight-hour periods (or shifts) in a working day, and is shut down for the remaining eight-hour shift.) There is also a drive to reduce the consumption of controllable make-up water, although this necessarily increases with the amount of load changing to which the boiler is subjected. A typical scenario is that if the boiler experiences a large load reduction (say 50%) the deaerator level would tend to rise because as the steam flow from the boiler reduces so must the feed flow. During this time, the condensate flowing to the deaerator would not be required by the feed pump, causing the level in the storage vessel to rise. The level controls would react by dumping the excess water that is filling the deaerator. If there is a load increase to the original load, the deaerator level would drop due to the increased flow from the feed pump, and the level controls would tend to add make-up water to the condenser. This cycle of making-up and spilling can occur every time the load changes.

Table 6.1 Operation of deaerator level-control valves
(© Scottish Power plc. (Reproduced by permission)

Controller output	Valve duty	Valve position
0–25%	Spill valve	100–0%
25–75%	Deadband	n/a
75–100%	Make-up valve	0–100%

The requirement to reduce the wasted make-up water has resulted in the use of advanced control strategies for controlling the make-up and spill on larger boilers. For example a fuzzy logic control algorithm embedded in the DCS software can be used. The inputs to the controller are the steam flow and the deaerator level, the outputs are to the make-up and spill valves. The system uses rule-based logic to take decisions on making-up and spilling. The goal of the scheme is to minimise make-up and spill by retaining water within the condensate system until it is required.

6.7 Summary

Once the combustion process has occurred, the feed water has boiled and the steam has been generated, the next requirement is to ensure that the temperature of the steam that is delivered to the turbine or heat load is maintained at the correct value. In the next chapter we shall look at the control and instrumentation systems that are employed for this purpose.

Chapter 7

Steam-temperature control

7.1 Why steam-temperature control is needed

The rate at which heat is transferred to the fluid in the tube banks of a boiler or HRSG will depend on the rate of heat input from the fuel or exhaust from the gas turbine. This heat will be used to convert water to steam and then to increase the temperature of the steam in the superheat stages. In a boiler, the temperature of the steam will also be affected by the pattern in which the burners are fired, since some banks of tubes pick up heat by direct radiation from the burners. In both types of plant the temperature of the steam will also be affected by the flow of fluid within the tubes, and by the way in which the hot gases circulate within the boiler.

As the steam flow increases, the temperature of the steam in the banks of tubes that are directly influenced by the radiant heat of combustion starts to decrease as the increasing flow of fluid takes away more of the heat that falls on the metal. Therefore the steam-temperature/steam-flow profile for this bank of tubes shows a decline as the steam flow increases.

On the other hand, the temperature of the steam in the banks of tubes in the convection passes tends to increase because of the higher heat transfer brought about by the increased flow of gases, so that this temperature/flow profile shows a rise in temperature as the flow increases. By combining these two characteristics, the one rising, the other falling, the boiler designer will aim to achieve a fairly flat temperature/flow characteristic over a wide range of steam flows.

No matter how successfully this target is attained, it cannot yield an absolutely flat temperature/flow characteristic. Without any additional control, the temperature of the steam leaving the final superheater of the boiler or HRSG would vary with the rate of steam flow, following what is known as the 'natural characteristic' of the boiler. The shape of this will depend on the particular design of plant, but in general, the temperature will rise to a peak as the load increases, after which it will fall.

The steam turbine or the process plant that is to receive the steam usually requires the temperature to remain at a precise value over the entire load range, and it is mainly for this reason that some dedicated means of regulating the temperature must be provided. Since different banks of tubes are affected in different ways by the radiation from the burners and the flow of hot gases, an additional requirement is to provide some means of adjusting the temperature of the steam within different parts of the circuit, to prevent any one section from becoming overheated.

In theory, the design of the plant should be targeted on arranging for the natural characteristic to attain the correct steam temperature when the rate of steam flow is that at which the boiler will normally operate. If this is possible, it means that spray water is used only while the unit is being brought up to load or when it operates at off-design conditions. In practice this objective can be attained only to a limited extent, because the boiler's natural characteristic changes with time due to factors such as fouling of the metal surfaces, which affects the heat transfer. In general, it is common to operate with continuous spraying, which has the advantage of allowing the steam temperature to be adjusted both upwards and downwards. If the required temperature were to be met solely by employing the natural characteristic as described, it would not be possible to produce temperature increases.

Before looking at the types of steam-temperature control systems that are applied, it will be useful to examine some of the mechanisms which are employed to regulate the temperature according to the controller's commands. Depending on whether or not the temperature of the steam is lowered to below the saturation point the controlling devices are known as attemperators or desuperheaters. (Strictly speaking, the correct term to use for a device which reduces the steam temperature to a point which is still above the saturation point is an attemperator, while one that lowers it below the saturation point may be referred to either as an attemperator or a desuperheater. However, in common engineering usage both terms are applied somewhat indiscriminately.)

7.2 The spray-water attemperator

One way of adjusting the temperature of steam is to pump a fine spray of comparatively cool water droplets into the vapour. With the resulting intermixing of hot steam and cold water the coolant eventually evaporates so that the final mixture comprises an increased volume of steam at a temperature which is lower than that prior to the water injection point. This cooling function is achieved in the attemperator.

The attemperator is an effective means of lowering the temperature of the steam, though in thermodynamic terms it results in a reduction in the performance of the plant because the steam temperature has to be raised to a higher value than is needed, only to be brought down to the correct value later, by injecting the spray water.

Although the inherent design of the attemperation system may, in theory, permit control to be achieved over a very wide range of steam flows, it should be understood that the curve of the boiler's natural characteristic will restrict the load range over which practical temperature control is possible, regardless of the type of attemperator in use. It is not unusual for the effective temperature-control range of a boiler to be between only 75% and 100% of the boiler's maximum continuous rating (MCR). This limitation is also the result of the spray-water flow being a larger proportion of the steam flow at low loads.

7.2.1 The mechanically atomised attemperator

Various forms of spray attemperator are employed. Figure 7.1 shows a simple design where the high-pressure cooling water is mechanically atomised into small droplets at a nozzle, thereby maximising the area of contact between the steam and the water. With this type of attemperator the water droplets leave the nozzle at a high velocity and therefore travel for some distance before they mix with the steam and are absorbed. To avoid stress-inducing impingement of cold droplets on hot pipework, the length of straight pipe in which this type of attemperator needs to be installed is quite long, typically 6 m or more.

With spray attemperators, the flow of cooling water is related to the flow rate and the temperature of the steam, and this leads to a further limitation of a fixed-nozzle attemperator. Successful break-up of the water into atomised droplets requires the spray water to be at a pressure which exceeds the steam pressure at the nozzle by a certain amount (typically 4 bar). Because the nozzle presents a fixed-area orifice to the spray water, the pressure/flow characteristic has a square-law shape, resulting in a restricted range of flows over which it can be used (this is referred to as limited turn-down or rangeability). The turn-down of the mechanically atomised type of attemperator is around 1.5 : 1.

The temperature of the steam is adjusted by modulating a separate spray-water control valve to admit more or less coolant into the steam.

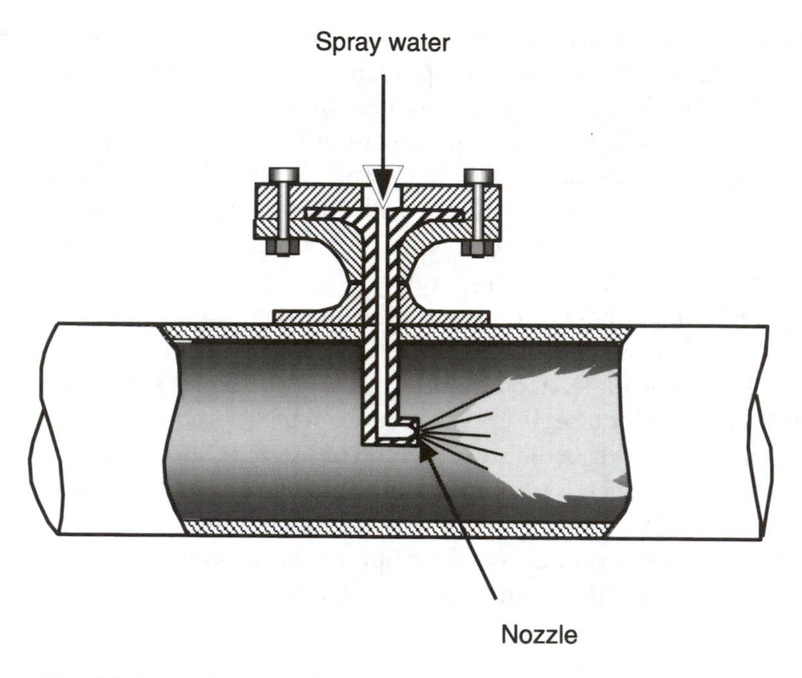

Figure 7.1 Mechanically atomised desuperheater

Because of the limitations of the single nozzle, the accuracy of control that is possible with this type of attemperator is no greater than $\pm 8.5\,^{\circ}$C.

7.2.2 The variable-area attemperator

One way of overcoming the limitations of a fixed nozzle in an attemperator is to use an arrangement which changes the profile as the throughput of spray water alters. Figure 7.2 shows the operating principle of a variable area, multinozzle attemperator. This employs a sliding plug which is moved by an actuator, allowing the water to be injected through a greater or smaller number of nozzles. With this type of device, the amount of water injected is regulated by the position of the sliding plug, a separate spray-water control valve is therefore not needed.

Adequate performance of this type of attemperator depends on the velocity of the vapour at the nozzles being high enough to ensure that the coolant droplets remain in suspension for long enough to ensure their absorption by the steam. For this reason, and also to provide thermal protection for the pipework in the vicinity of the nozzles, a thermal liner is often included in the pipe extending from the plane of the nozzles to a point some distance downstream.

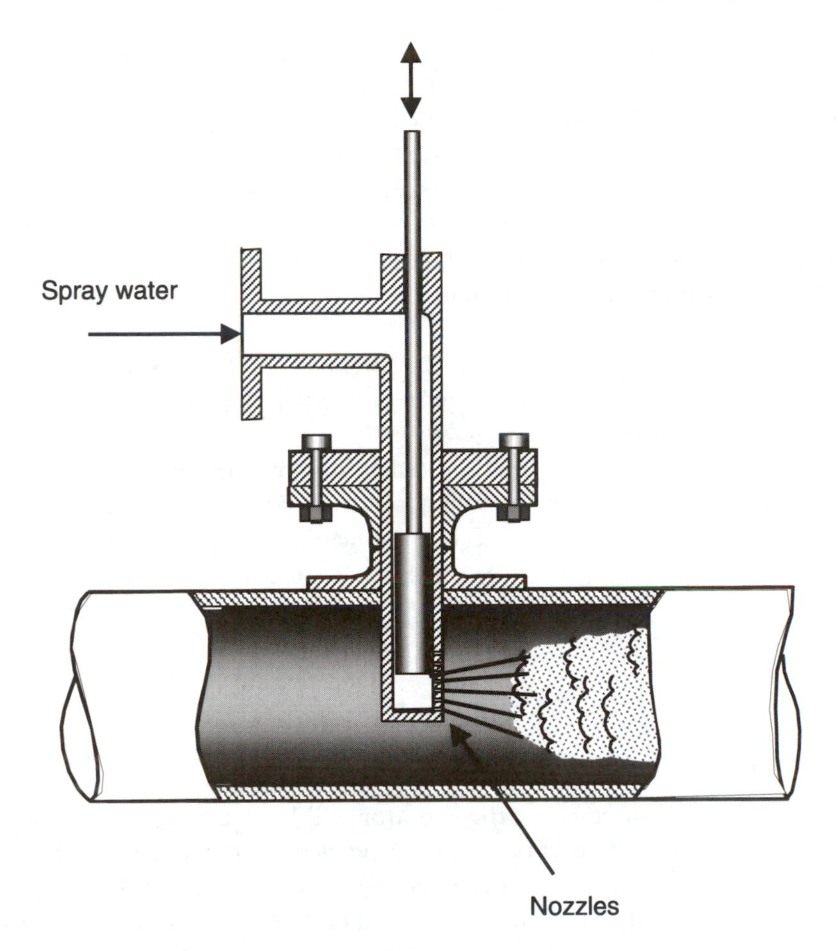

Figure 7.2 Principle of a multinozzle desuperheater

The accuracy of control and the turndown range available from a multi-nozzle attemperator is considerably greater than that of a single-nozzle version, allowing the steam temperature to be controlled to $\pm 5.5°C$ over a flow range of 40 : 1.

7.2.3 The variable-annulus desuperheater

Another way of achieving accurate control of the steam temperature over the widest possible dynamic range is provided by the variable-annulus desuperheater (VAD) (produced by Copes-Vulcan Limited, Road Two, Winsford Industrial Estate, Winsford, Cheshire, CW7 3QL.). Here, the approach contour of the VAD head is such that when the inlet steam flows through an annular ring between the spray head and the inner wall of the steam pipe its velocity is increased and the pressure slightly reduced. The

coolant enters at this point and undergoes an instant increase in velocity and a decrease in pressure, causing it to vapourise into a micron-thin layer which is stripped off the edge of the spray head and propelled downstream. The stripping action acts as a barrier which prevents the coolant from impinging on the inner wall of the steam pipe. The downstream portion of the VAD head is contoured, creating a vortex zone into which any unabsorbed coolant is drawn, exposing it to a zone
of low pressure and high turbulence, which therefore causes additional evaporation.

Due to the Venturi principle, the pressure of the cooled steam is quickly restored downstream of the vena contracta point, resulting in a very low overall loss of pressure.

An advantage of the VAD is that, due to the coolant injection occurring at a point where the steam pressure is lowered, the pressure of the spray water does not have to be significantly higher than that of the steam.

7.2.4 Other types of attemperator

At least two other designs of attemperator will be encountered in power-station applications. The vapour-atomising design mixes steam with the cooling water, thus ensuring more effective break-up of the water droplets and shrouding the atomised droplets in a sheath of steam to provide rapid attemperation.

Variable-orifice attemperators include a freely floating plug which is positioned above a fixed seat—a design that generates high turbulence and more efficient attemperation. The coolant velocity increases simultaneously with the pressure drop, instantly vaporising the liquid. Because of the movement of the plug, the pressure drop across the nozzle remains constant (at about 0.2 bar). The design of this type of attemperator is so efficient that complete mixing of the coolant and the steam is provided within 3 to 4 m of the coolant entry point, and the temperature can be controlled to $\pm 2.5\,^{\circ}$C, theoretically over a turndown range of 100 : 1.

Because the floating plug moves against gravity, this type of attemperator must be installed in a vertical section of pipe with the steam through it travelling in an upward direction. However, because of the efficient mixing of steam and coolant, it is permissible to provide a bend almost immediately after the device. Figure 7.3 shows a typical installation.

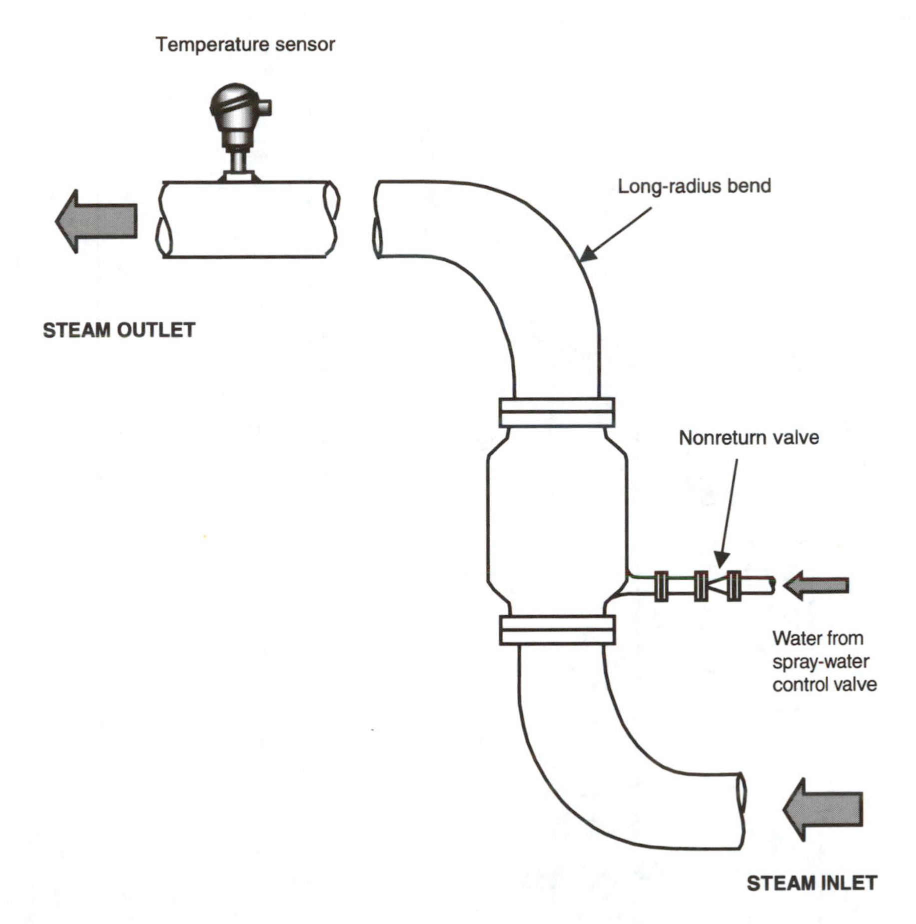

Figure 7.3 Variable-orifice attemperator installation

7.2.5 Location of temperature sensors

Because the steam and water do not mix immediately at the plane of the nozzle or nozzles, great care must be taken to locate the temperature sensor far enough downstream of the attemperator for the measurement to accurately represent the actual temperature of the steam entering the next stage of tube banks. Direct impingement of spray water on the temperature sensor will result in the final steam temperature being higher than desired. Figure 7.4 shows a typical installation, in this case for a variable-annulus desuperheater.

7.2.6 Control systems for spray-water attemperators

The simplest possible type of control would be based on measuring the temperature of the steam leaving the final superheater, and modulating

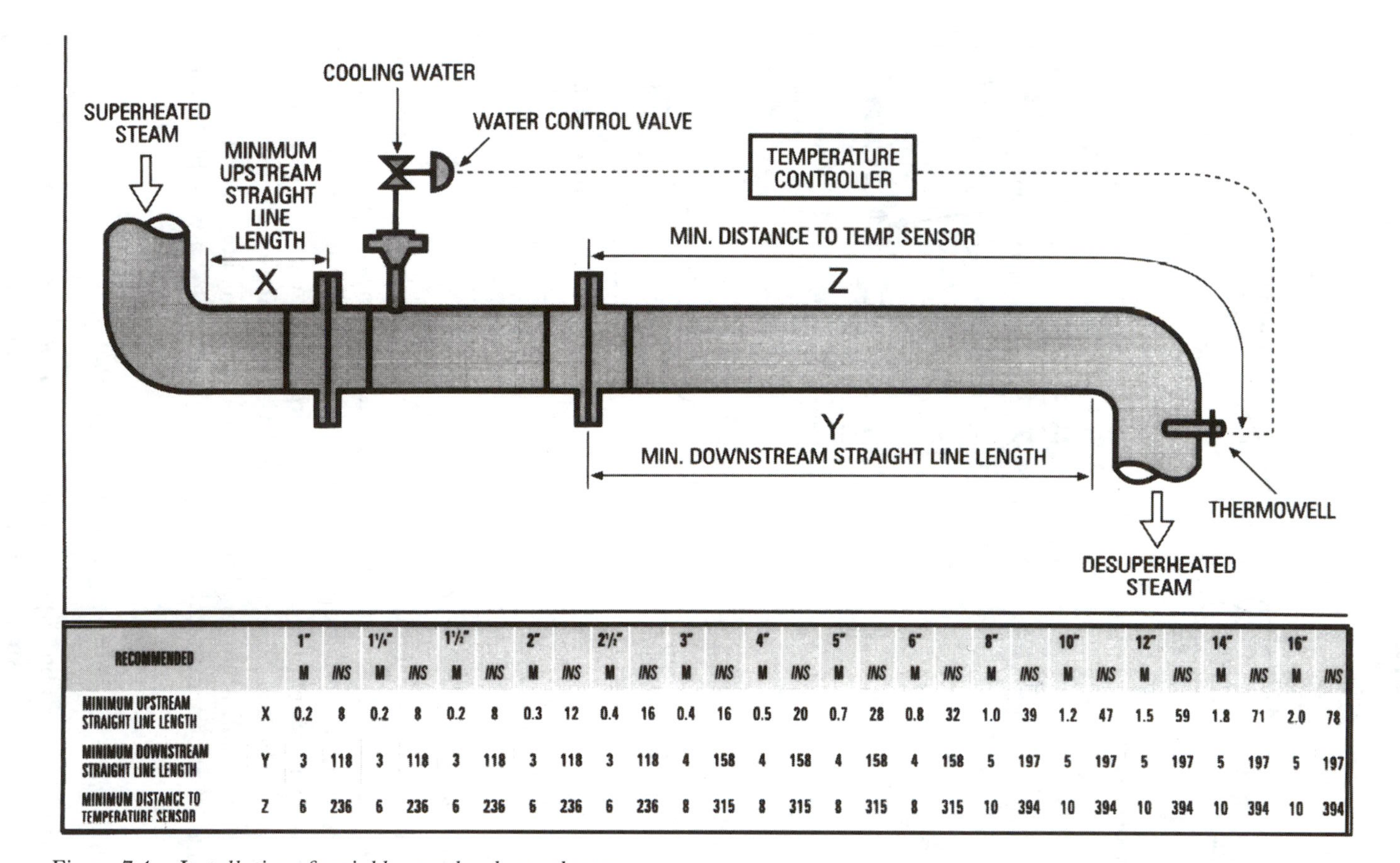

RECOMMENDED		1"		1¼"		1½"		2"		2½"		3"		4"		5"		6"		8"		10"		12"		14"		16"	
		M	INS	M	INS	M	INS	M	INS	M	INS	M	INS	M	INS	M	INS	M	INS	M	INS	M	INS	M	INS	M	INS	M	INS
MINIMUM UPSTREAM STRAIGHT LINE LENGTH	X	0.2	8	0.2	8	0.2	8	0.3	12	0.4	16	0.4	16	0.5	20	0.7	28	0.8	32	1.0	39	1.2	47	1.5	59	1.8	71	2.0	78
MINIMUM DOWNSTREAM STRAIGHT LINE LENGTH	Y	3	118	3	118	3	118	3	118	3	118	4	158	4	158	4	158	4	158	5	197	5	197	5	197	5	197	5	197
MINIMUM DISTANCE TO TEMPERATURE SENSOR	Z	6	236	6	236	6	236	6	236	6	236	8	315	8	315	8	315	8	315	10	394	10	394	10	394	10	394	10	394

Figure 7.4 Installation of variable-annulus desuperheater

the flow of cooling water to the spray attemperator so as to keep the temperature constant at all flow conditions. Unfortunately, because of the long time constants associated with the superheater, this form of control would produce excessive deviations in temperature, and a more complex arrangement is required.

Two time constants are associated with the superheater. One represents the time taken for changes in the firing rate to affect the steam temperature, the other is the time taken for the steam and water mixture leaving the attemperator to appear at the outlet of the final superheater. In terms of temperature control it is the latter effect which predominates because, although changes in heat input will affect the temperature of the steam, a fast-responding temperature-control loop will be able to compensate for the alterations and keep the temperature constant. It is the reaction time between a change occurring in the spray-water flow and the effects being observed in the final temperature that determines the extent of the temperature variations that will occur.

Another problem with a simple system, as outlined above, is that it does not permit any monitoring and control of the temperature to occur *within* the steam circuit—only at the exit from the boiler.

These difficulties are addressed by the application of a cascade control system as shown in Figure 7.5. This shows a simple steam-temperature control system based on the use of an interstage attemperator which is located in the steam circuit between the primary and secondary banks of superheater tubes.

Since it is the temperature of the steam leaving the secondary superheater that is important, this parameter is measured and a corresponding signal fed to a two- or three-term controller (proportional-plus-integral or proportional-plus-integral-plus-derivative). In this controller the measured-value signal is compared with a fixed desired-value signal and the controller' s output forms the desired-value input for a secondary controller. (Because the output from one controller 'cascades' into the input of another, this type of control system is commonly termed 'cascade control'.)

The secondary controller compares this desired-value signal with a measurement representing the temperature of the steam immediately after the spray-water attemperator.

It is a matter of some debate as to whether a two-term or three-term controller should be used in this type of application. Because the steam-temperature sensors used are subjected to the high pressures and temperatures of the superheater, they have to be enclosed in substantial steel pockets. Even with the best designs, pockets are usually slow-responding, with the result that any high-speed fluctuations in the measured-value

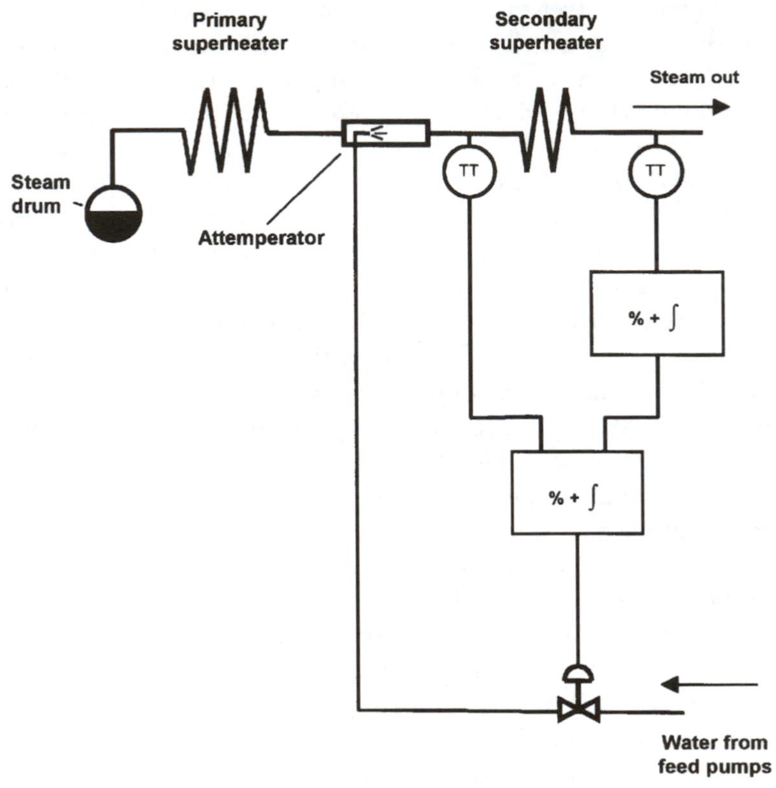

Figure 7.5 Steam-temperature control with a single interstage spray attemperator

signal will be smoothed out and the resultant signal will be fairly stable. The use of a derivative term is therefore easier than in, say, flow measurement applications where small-scale but sudden changes in flow can occur. When rapid input changes are differentiated, the controller output changes by a large amount, and for this reason tuning three-term flow controllers for optimum response can become difficult. This is not a problem with the temperature controllers described here, and the application of derivative action may be viable if it is felt that this could provide improved performance. As usual, it is important that the controller design should be such that the derivative term affects only the measured-value signal (not the desired-value or error signals), since differential response to operator-induced setpoint changes is always undesirable.

In Chapter 6, reference was made to the requirement for the spray water to enter the attemperator at a pressure which exceeds the steam pressure by a minimum value. It is worth remembering the point made there: that, where a variable-speed feed pump is used, care must be taken to ensure that adequate water/steam differential pressure is available under all operational conditions.

7.2.6.1 Controller saturation effects

The type of control system described above is commonly encountered in a wide variety of applications, and it is subject to an effect which must be understood and adequately addressed by the design of the controllers used in the system. The effect is known as 'integral saturation' or 'reset wind-up', and it is a characteristic of integral-action controllers whose output commands are fed into the inputs of cascade or secondary controllers. It sometimes confuses people when they are first introduced to this saturation effect in steam-temperature control applications since the word 'saturation' is also applied to a thermodynamic property of steam. It is therefore important that the point is clearly understood that in this context the word 'saturation' refers to a controller output reaching a limiting value and then attempting to exceed that figure.

Whether the implementation of a controller is achieved in hardware or by software its output must always be constrained by some design limit or other. At first sight, it may appear that the exception to this rule is the so-called pulse or incremental controller, where the output commands dictate a *change* in the position of the controlled device. However, if the controller and actuator are considered together as a system it will be seen that saturation will still occur when the actuator reaches the limit of its movement.

With a gain of x, the output of a controller will be the input error multiplied by x, but what happens if either x or the error is so large that the resulting output is outside the range of signals that the controller can handle (i.e. beyond the limiting value of the output range)? In this case the controller output will adopt a magnitude which is fixed at the limiting value, in which condition the output is no longer representative of the input error.

The range of inputs within which the controller output is representative of the input error is inversely proportional to the gain: with a large gain, a small input error may force the output signal to the limiting value, and vice versa.

It is for this reason that control engineers refer to the 'proportional band' of the controller, the proportional band of a controller being the reciprocal of its gain, with a gain of 1 being equivalent to a proportional band of 100%. It is worth exploring the significance of this statement with the following example.

In Figure 7.6*a* the controller has a gain of 1. The input error is defined as being in the range 0–100% and the output is also in the range 0–100%.

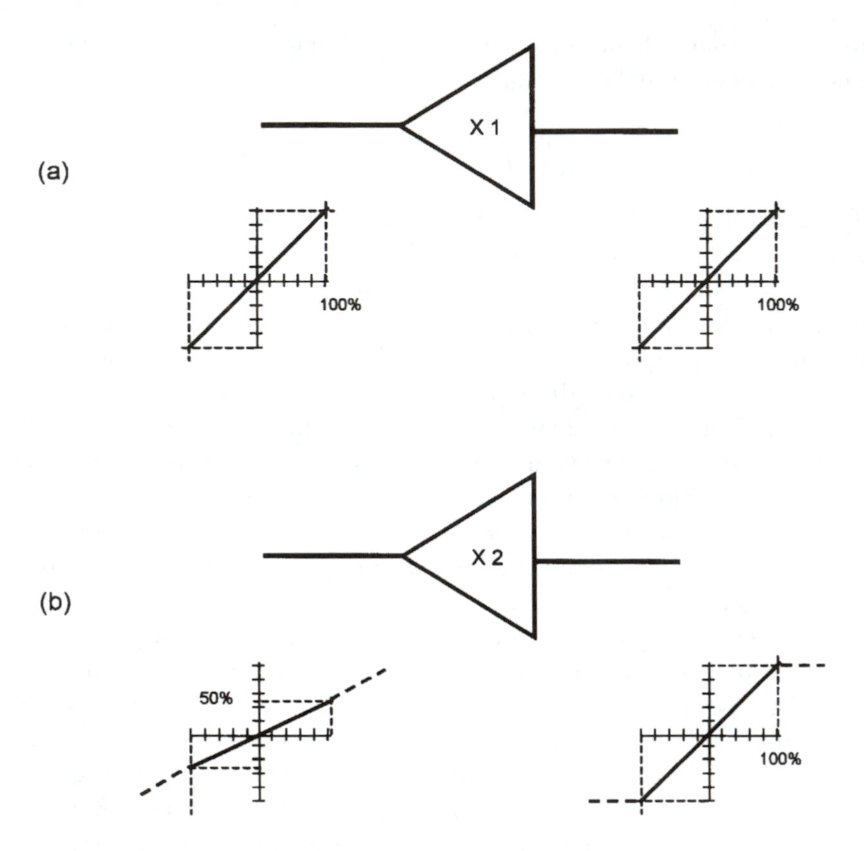

Figure 7.6 Controller saturation

If the input error is slowly swung from one extreme limit to the other (i.e. from −100% to +100%), the output also moves from one limit to the other (in the above case from −100% to +100%).

In Figure 7.6*b* the controller gain has been increased to 2, and this time the output reaches its limiting value of +100% when the input error is only +50%, and it becomes −100% when the input is −50%. In other words, the output moves from one limit to the other with an input swing of 50% of the maximum range. If the input is increased beyond the value of 50% in either direction, as shown by the dotted lines, the output cannot respond, since the output has reached the limit of its range and the controller has become saturated.

The significance of the expression 'proportional band' should now be apparent. It is the range of input signals within which the output is proportional to the input.

A good way of understanding what is happening is to think of the input in terms of a window. Every change that occurs within the window will be reacted to, but any change that occurs when the input is outside of the window is invisible to the controller. The size of the window is inversely proportional to the controller gain: the higher the gain, the smaller the window.

With a simple loop, the effects of controller saturation are not too severe since the output has reached the limit of movement of the actuator and it starts to react as soon as the controller input error has re-entered the proportional-band window. In a cascade loop, however, the presence of the two controllers leads to highly undesirable results.

To illustrate this point, Figure 7.7 shows a cascade loop with practical values of temperature and gain added. As is common with this type of system, the temperature transmitter has a suppressed range, with 4 mA being transmitted when the temperature is 300 °C and 20 mA when the temperature is 600 °C. When the steam temperature is 400 °C the error

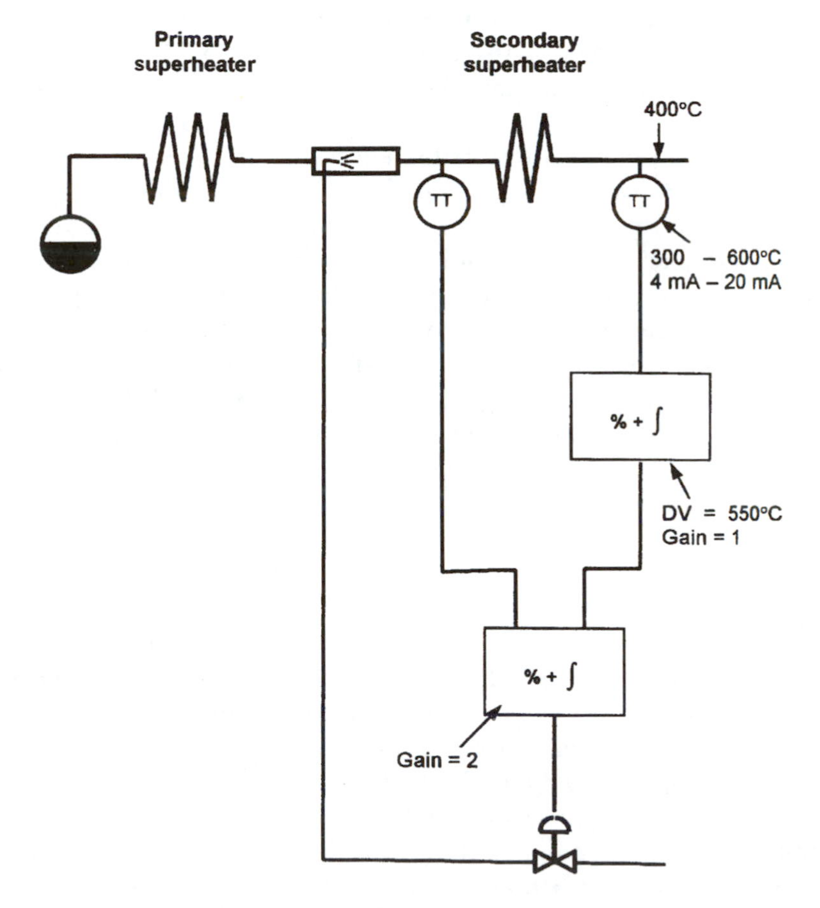

Figure 7.7 Saturation effects in steam-temperature control system

between the measured and desired value signals is 150 °C, which is 50% of the 300 °C input range and with a gain of 1 the controller output is also 50%. Since the secondary controller has a gain of 2, the 50% signal from the primary controller is right at the edge of its own proportional-band window.

Now let us see what happens when the temperature is outside the proportional-band window when, as stated above, the secondary controller cannot react to any changes in it. If the temperature is at, say, 350 °C and rising, the primary controller will react to the rise in temperature and produce output commands which should start increasing the flow of cooling spray-water. However, these commands are initially invisible to the secondary controller, and the spray valve will not be moved until the signal eventually enters the proportional-band window of the secondary controller. By then it is far too late, and the situation is exacerbated by the long time constant of the final superheater (up to $1\frac{1}{2}$ minutes in some cases). As a result, the temperature continues to rise, and when the spray valve finally opens the result is a severe overcorrection. The result is that the final steam temperature and the spray valve opening will both begin to oscillate.

This is a classical example of the effects of controller saturation, but the problem is not always understood by DCS vendors who are unfamiliar with boiler control systems. On the other hand, vendors who do recognise the problem offer a variety of solutions. These usually involve the use of 'track' and 'reset' facilities in the controllers (or the software function blocks which assume the duty of controllers). With such solutions the controller output is forced to follow a signal which is connected to its 'track' terminal when the reset (integration) function is disabled. In one configuration, the system monitors the output signal of the primary controller, and when this reaches 100% the controller output is forced to track the input temperature.

7.2.6.2 Prevention of over-cooling

In steam-temperature control applications it is important to prevent the temperature being reduced too far. If the temperature at the inlet of the secondary superheater falls to a value approaching the saturation temperature, water droplets could form in the flowstream, raising the possibility of thermal shock to the pipework, and in addition the steam circuit could become partially plugged. The flow through the obstructed tubes will then be reduced and their surface temperature will rise, possibly causing premature tube failure.

7.2.6.3 Multistage attemperators

Some boilers have several banks of superheater tubes. In these cases spray attemperators are normally provided between the major banks, as shown in Figure 7.8.

It will be seen that the control systems around each superheater comprise cascade loops that are quite similar to those discussed earlier. However, the set-value signal for the first stage of spraying is derived from the output of the controller regulating the final steam temperature. In fact, the signal may be characterised in some way to accurately represent the relationship between the temperature of the steam leaving the second stage of attemperation and that at the exit of the first stage. When the system is operating correctly, with the final slave controller maintaining its desired-value and measured-value signals at the same value, the effect is to maintain a constant temperature-differential across the second attemperator. The temperature drop across the attemperator is a measure of the

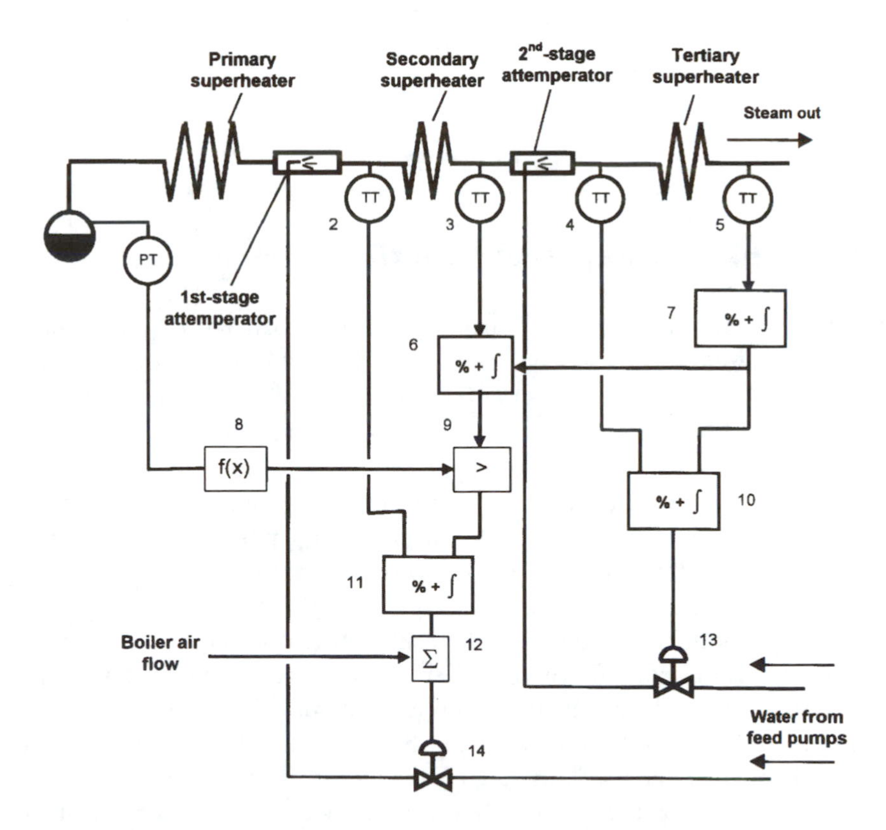

Figure 7.8 Steam-temperature control with two-stage spray attemperation

work being done by it, and by controlling this to a known value the cooling load through the entire string of tubes can be apportioned as required.

The function of the maximum-selector unit (item 9) is to prevent chilling as described in Section 7.2.6.2. The steam pressure at the drum is measured and its value characterised (8) to produce a signal which represents the saturation temperature. Item 8 also incorporates a bias to represent a safe margin of operation and the resultant signal is fed to the maximum-selector unit. If the output of the secondary-superheater controller should fall to a value that is at or below the safety margin above the saturation temperature, it is ignored by the maximum-selector unit, which clamps the desired value signal for the first stage attemperator at this limit.

Another feature of the system shown in this diagram is the programming signal which presets the opening of the first-stage spray-water valve according to a characteristic that the boiler designer has predicted. The temperature-control systems then trim this opening to eliminate any residual error. These programming signals can overcome some of the boiler's time-delays, producing a better and faster response to changes in load.

7.3 Temperature control with tilting burners

As explained in Chapter 3, the burning fuel in a corner-fired boiler forms a large swirling fireball which can be moved to a higher or lower level in the furnace by tilting the burners upwards or downwards with respect to a midposition. The repositioning of the fireball changes the pattern of heat transfer to the various banks of superheater tubes and this provides an efficient method of controlling the steam temperature, since it enables the use of spray water to be reserved for fine-tuning purposes and for emergencies. In addition, the tilting process provides a method of controlling furnace exit temperatures.

With such boilers, the steam temperature control systems become significantly different from those of boilers with fixed burners. The boiler designer is able to define the optimum angular position of the burners for all loads, and the control engineer can then use a function generator to set the angle of tilt over the load range to match this characteristic. A temperature controller trims the degree of tilt so that the correct steam temperature is attained.

7.4 Controlling the temperature of reheated steam

In boilers with reheat stages, changes in firing inevitably affect the temperature of both the reheater and the superheater. If a single control mechanism were to be used for both temperatures the resulting interactions would make control-system tuning difficult, if not impossible, to optimise. Such boilers therefore use two or more methods of control.

Because of the lower operating pressure of reheat steam systems, the thermodynamic conditions are significantly different from those of superheaters, and the injection of spray water into the reheater system has an undue effect on the efficiency of the plant. For this reason, it is preferable for the reheat stages to be controlled by tilting burners (if these are available) or by apportioning the flow of hot combustion gases over the various tube banks. However, if the superheat temperature is controlled by burner tilting, gas apportioning or spray attemperation must then be used for the reheat stages.

In boilers with fixed burners, steam-temperature control may be achieved by adjusting the opening of dampers that control the flow of the furnace gases across the various tube banks. In some cases two separate sets of dampers are provided: one regulating the flow over the superheater banks, the other controlling the flow over the reheater banks.

Between them, these two sets of dampers deal with the entire volume of combustion gases passing from the furnace to the chimney. If both were to be closed at the same time, the flow of these gases would be severely restricted, leading to the possibility of damage to the structure due to overpressurisation. For this reason the two sets are controlled in a so-called 'split-range' fashion, with one set being allowed to close only when the other has fully opened.

These dampers provide the main form of control, but the response of the system is very slow, particularly with large boilers, where the temperature response to changes in heat input exhibits a second-order lag of almost two minutes' duration. For this reason, and also to provide a means of reducing the temperature of the reheat steam in the event of a failure in the damper systems, spray attemperation is provided for emergency cooling.

The spray attemperator is shut unless the temperature at the reheater outlet reaches a predetermined high limit. When this limit is exceeded, the spray valve is opened. In this condition, the amount of water that is injected is typically controlled in relation to the temperature at the reheater inlet, to bring the exit temperature back into the region where gas-apportioning or burner tilting can once again be effective. The relationship between the cold reheat temperature and the required spray-water flow can be defined by the boiler designer or process engineer.

If a turbine trip occurs the reheat flow will collapse. In this situation the reheat sprays must be shut immediately in order to prevent serious damage being caused by the admission of cold spray water to the turbine.

7.4.1 Spray attemperators for reheat applications

At first, it may seem that reheat spray-water attemperator systems should be similar to those of the superheater. This is untrue, because reheat attemperators have to cope with the lower steam pressure in this section of the boiler, which renders the pressure of the water at the discharge of the feed pumps too high for satisfactory operation. Although a pressure-reducing valve could be introduced into the spray-water line, this would be an expensive solution whose long-term reliability would not be satisfactory because of the severe conditions to which such a valve would be subjected. A better solution would be to derive the supply from the feed-pump inlet. In some cases, even this is ineffective, and separate pump sets have to be provided for the reheat sprays.

7.5 Gas recycling

Where boilers are designed for burning oil, or oil and coal in combination, they are frequently provided with gas-recirculation systems, where the hot gases exiting the later stages of the boiler are recirculated to the bottom part of the furnace, close to the burners. This procedure increases the mass-flow of gas over the tube banks, and therefore increases the heat transfer to them.

Because the gas exiting the furnace is at a low pressure, fans have to be provided to ensure that the gas flows in the correct direction. Controlling the flow of recycled gases provides a method of regulating the temperature of the superheated and reheated steam, but interlocks have to be provided to protect the fan against high-temperature gases flowing in a reverse direction from the burner area if the fan is stopped or if it trips.

7.6 Summary

This ends our system-by-system survey of boiler and HRSG control and instrumentation systems, and we will now turn our attention to some of the design aspects relating to the equipment that is used in implementing these systems.

Chapter 8

Control equipment practice

On an operational plant, the control systems that have so far been examined may be implemented in any of a variety of ways, ranging from pneumatics to advanced computer-based systems, but in all cases it should be possible to identify the various loops within the relevant configuration. These days, most control functions are implemented by means of a computer-based system, so we shall now briefly look at a typical configuration. After that, we shall examine some of the other hardware used in the systems and then consider the environmental factors that influence the selection of control and instrumentation equipment.

8.1 A typical DCS configuration

DCS stands for 'distributed control system'. The term 'distributed' means that several processors are operating together. This is usually achieved by dedicating tasks to different machines. It does not necessarily mean that the separate computers are physically located in different areas of the plant.

Figure 8.1 shows how a typical system may be arranged. The following notes relate to individual parts of that system. In practice, each manufacturer will usually offer some variant of the system shown in this diagram, and the relevant description should be consulted, but the comments made here are general ones which may help to identify points which should be considered and discussed when a new or refurbished system is being considered.

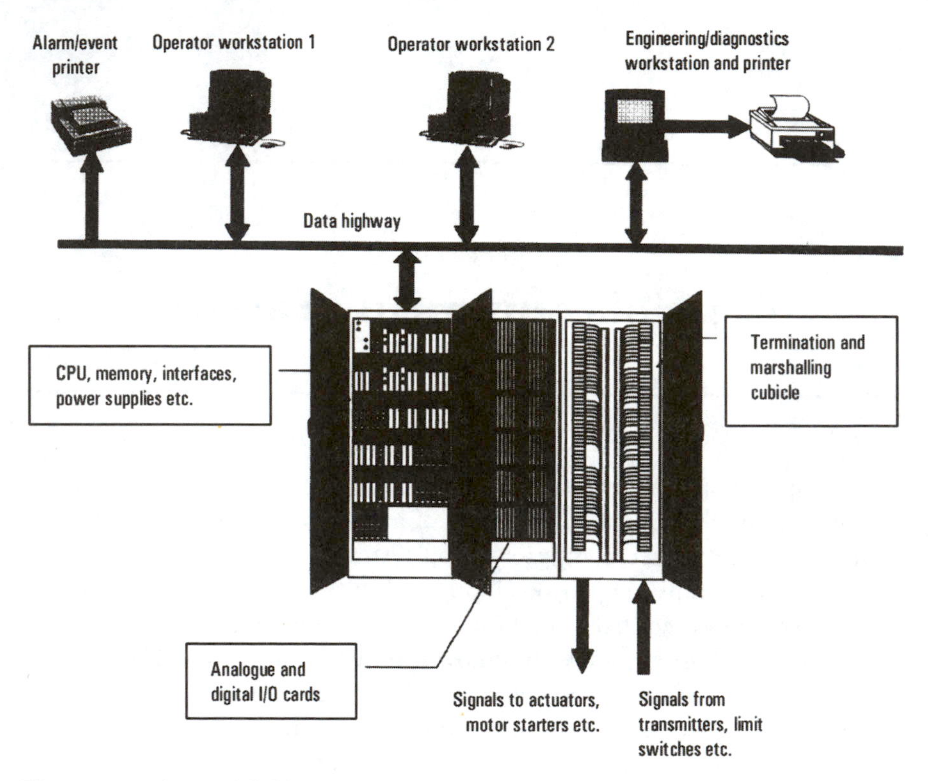

Figure 8.1 A typical DCS system configuration

8.1.1 The central system cabinets

Located near the centre of Figure 8.1 are the cabinets which house the processors that execute the control functions. These cubicles also contain the attendant interface and input/output (I/O) cards and the necessary power supply units (PSUs). The latter will usually be duplicated or triplicated, with automatic changeover from one to another in the event of the first failing. (This automatic changeover is often referred to as 'diode auctioneering' because silicon diodes are used to feed power from each unit onto a common bus-main. In the event of the operational power-supply unit failing, its diode prevents a power reversal while the back-up power unit takes over.) At this time it is important that the system should raise an alarm to warn that a PSU failure has occurred. Otherwise the DCS will continue to operate with a diminished power-supply reserve and any further failure could have serious consequences.

Clearly, the DCS cannot operate continuously from batteries alone. A reliable and stable source of power will therefore need to be available (usually 110 V or 240 V AC). If the DCS includes internal back-up batteries it will continue to operate if the AC feed is lost, but such batteries are normally sized to retain essential data in the memory and to provide a limited amount of functionality. They may also allow limited control to be performed, but all this will function for only a short period (typically 30 min) and it is therefore usual to provide an external uninterruptable power supply (UPS) system which can allow the plant to be operated for a longer time. The duration of this period warrants very careful consideration. Long periods require large and expensive batteries and charger systems, and this expense can rarely be justified (especially since such a major power loss will probably have disabled all pumps, motors etc.). Instead, it is common to provide a battery capacity that will allow the plant to be safely shut down in the event of power failure. The determination of the time required for such an operation is a matter of discussion with the process design engineers and the plant management.

In addition to supplying the computer system, the power-supply system will usually also have to provide DC supplies for 4–20 mA transmitters and for limit-switch contacts. (The voltage connected to a contact and thence to the DCS input channel is often referred to as the 'wetting voltage'.) Transmitters operating on the 4–20 mA range which are powered from the DCS are sometimes called 'passive'. In comparison, those that operate from local power supplies are called 'active'.

The I/O cards consist of analogue and digital input and output channels. Analogue inputs convert the incoming 4–20 mA signals to a form which can be read by the system. The printed-circuit cards for analogue inputs may or may not provide 'galvanic isolation'. With a galvanically isolated device the signal circuit is electrically isolated from others, from the system earth and from the power-supply common rail. Galvanic isolation simplifies circuit design since it prevents inadvertent short-circuiting, but consideration should be given to the possible build-up of static charges on completely ungrounded circuits, which could cause damage to input devices (which are usually rated for not more than a few tens or hundreds of volts). This is normally an important consideration only in areas of very low humidity or where there is a strong presence of charged particles.

The commissioning process, and the task of identifying and correcting faults, are operations which are considerably assisted by the provision of light-emitting diode status indicators (LEDs) on the digital output cards.

Some systems provide switches on the digital input cards, which can be of assistance with commissioning and fault-finding. However, inadvertent or deliberate maloperation of such switches can have serious consequences, since the DCS is then provided with incorrect plant-status information and it may take inappropriate action. (The use of logic probes, which inject signals into a system to check its operation, is also to be deprecated, for similar reasons.)

Analogue and digital I/O channels are normally grouped into 8, 16, 32 or 64 channels per printed-circuit card. 8 or 16 analogue input (AI) channels are commonly accommodated on a card, but analogue output (AO) channels consist of current generators and so occupy more space and are more expensive than AI channels, which are based on small operational-amplifier devices (op-amps). Digital input (DI) channels are very simple and cheap and may be grouped into 16 or even 32 inputs to a single card. Digital output (DO) channels driving lower-power devices are also simple and cheap, and may also comprise 16 or 32 inputs to a single card, but DOs for higher-power devices (such as solenoid valves) usually require the provision of relays. These may be included on the card or they may be separate.

When considering the provision of spare I/O channels, careful thought must be given to the grouping of channels. If a system has 256 analogue input channels available, of which only 230 are actively used, it may be said to have 11% spare capacity in this area. However, the grouping of functional areas into cards will inevitably result in the occurrence of more spare channels in one area than in another. It is possible, therefore, to have the required amount of spare I/O capacity available in terms of the overall system, but to be unable to modify or extend a particular part of the system safely, because no spare channels have been provided in the required area.

Spare capacity should be provided both in the form of 'populated' channels (i.e. spare inputs and outputs on individual cards) and 'unpopulated' space (i.e. spaces for additional cards). To avoid a spaghetti-like tangle of cross-connections, the spare spaces should be sensibly distributed through the system.

8.1.2 Termination and marshalling

It is important to understand that the grouping of inputs and outputs on the I/O cards does not always correspond with the grouping of signals into multipair cables, which is dictated by the physical arrangement of equipment on the plant. While it is sensible to avoid mixing different control systems (e.g. feed water control and combustion control) onto a single card, the signals associated with a single system will not necessarily all be carried in the same cable. The result is that a certain degree of cross-connection or 'marshalling' is always required.

Well-designed systems will provide adequate facilities for neatly marshalling the signal connections, but this inevitably requires that the identification of signal connections and their location in the cable system is known at an early stage of the contract. The later this problem is resolved, the more complex and untidy the system will become. Complexity and untidiness can be dangerous because it can lead to mistakes occurring during commissioning or afterwards.

8.1.3 Operator workstations

The operator workstations consist of screens on which plant information can be observed, plus keyboards, trackballs or 'mouse' devices allowing the operator to send commands to the system. They also comprise printers for operational records, logging of events (such as start-up of a pump), or alarms. Some systems also provide plotters (one use of plotters is to detect the possible stalling of an axial-flow fan, as described in Chapter 3).

The screens can be ordinary cathode-ray tube types as used with personal computers, or they may be large-screen plasma displays or projection systems. The selection of the type of screen depends on the operational requirements, but will ultimately be determined by the available budget. Critical ergonomic factors affect the optimum design of the workstations, and great care must be exercised to ensure that the plant can be operated safely under all conceivable modes of failure, and that no computer-assisted errors can occur due to the operator being confused by the information presented to him or her.

An important consideration is the screen update time. This is the time between the occurrence of an event and its appearance on the screen. As system loading is increased, this time can become extended, but the operator will need to be made aware of each event as soon as possible after it occurs, so that corrective action can be taken. An update time of 1 s is barely adequate to deal with fast-moving events, but it can be quite difficult to achieve.

8.2 Interconnections between the systems

The considerations applying to field cabling are dealt with in Section 8.8. However, special thought needs to be given to the data highway. This is a high-speed link over which a great deal of information is transmitted. The cable employed for this purpose is very specialised, and great care has to be taken in its installation. Physical damage, severe bending or incorrect termination can cause maloperation. If a fibre-optic cable is used, the considerations that apply to this type of cabling must be meticulously followed.

The integrity of the data highway is crucial to the safety of the plant and therefore it is usually duplicated. However, the provision of a sophisticated dual-redundant highway with full error-checking and correction has on occasion been completely negated by the cable installer running both cables on the same tray, or over the same route. An incident that damages one cable will in all probability also damage the second one, with severe consequences.

8.3 Equipment selection and environment

Although modern gas-fired plant naturally tends to be clean in comparison with its coal-fired equivalents, any power-station environment still presents a severe test for electronic systems. The control-system designer has to deal with the problems of operating low-voltage, potentially interference-prone, electronic equipment in close proximity to electrical plant operating at 11 kV and above, with all its attendant switchgear and transformers. The situation is exacerbated when considerations of safe operation in hazardous environments are brought into the picture. It becomes even worse when considering the dust, dirt and vibration that are significant factors in practical power-plant environments. Naturally, the latter problems (dust and dirt) become particularly acute in coal-fired plant.

The success of a control system depends on the designer understanding and addressing these factors. To assist in this process the following chapter provides an outline of good equipment design and installation practices. Because the subject covers so many different disciplines, the chapter is divided into three sections:

- Mechanical factors: the ground rules for providing good facilities for control and instrumentation equipment.
- Electromagnetic compatibility: guidelines for minimising the risk of maloperation caused by interference.
- Physical environmental considerations: dealing with dust, dirt, vibration and hazardous atmospheres.

These matters must be understood and judiciously applied when an installation is being planned, but doing this involves considerable interplay with the civil and mechanical-engineering disciplines, and appropriate action must therefore be taken at a very early stage in the design and construction phases of the plant. In a new plant, given diligence and understanding on the part of all the disciplines involved, one can hope to achieve this goal. But in the case of a refurbishment project the task becomes much more difficult, because here one is dealing with a plant whose construction is already complete. In this case the control-system designer must work with what already exists. In the end it may come to a matter of fighting a ditch-by-ditch battle, eventually retreating to the last principle—the one that must never be sacrificed—which is to obtain an installation that is safe to operate and maintain.

8.4 Mechanical factors and ergonomics

In this section we shall consider the mechanical installation of electronic control equipment. This is necessarily a summary, and like most aspects of technology it is affected by changing requirements and technologies. Because requirements, technologies and the availability of materials are always changing, some form of guidance on up-to-date practice should be sought and it is tempting to think that the selected system vendor will be able to provide this.

Reputable vendors should be pleased to provide guidance on installation practices to be employed with their equipment and systems. This is partly because by providing such information they demonstrate that they are experienced in power-station work, and are able and willing to help with such matters. It is also in their own interest to do everything possible to ensure that their systems will not be exposed to mechanical or electromagnetic conditions that could jeopardise their performance.

However, it will be unwise to wait until a specific vendor has been selected because in most cases this action occurs when the basic concrete and steel construction is almost complete, by which time it will be too late to make any changes. It is useful to obtain guidance on current practice from a range of system vendors.

8.4.1 Site considerations

The electronic assemblies that comprise a control system will generally be located in three areas:

- The field: where transmitters, sensors, detectors and actuators are sited.
- The equipment room(s): accommodating the control cubicles, processors, I/O facilities and power-supplies.
- The control room: housing the operator facilities (screens and keyboards), plus the system printers etc.

These areas represent very different environments for the equipment they contain, ranging from the severe conditions of dust, dirt, humidity, heat, vibration and hazardous areas that are to be found on the plant, to the comparatively quiet and clean conditions that should be found in the control room.

8.4.1.1 Field equipment

Every control system depends for its operation on accurate information on the plant being controlled, (which is the duty of the process transmitters), and the ability to apply the resulting commands to the plant (which is done by actuators). These electromechanical transducers are vital to the proper operation of the system. They *must* operate efficiently and reliably.

Unfortunately, items of equipment in these critical areas are subject to particularly severe difficulties. For a start, the design of electromechanical transducers requires a blend of good electronic engineering and mechanical engineering. A thorough understanding of metallurgy and engineering chemistry is often also required. Designing equipment where several disciplines are involved is much more difficult than working in only one discipline. If this hurdle is overcome successfully a good device will result, but it will then be installed on the plant where it will be exposed to the severe environments that often exist there. The equipment's operation will then depend heavily on the application of the best possible installation and maintenance practices.

A successful control system requires detailed definition of each component part, and in the case of the transmitters this is achieved by meticulous specification of the transmitter itself, and by careful definition of how it will be installed.

Figure 8.2 shows one step in the latter process, a so-called 'hook-up' diagram for a transmitter measuring high pressures. The hook-up defines how the transmitter is piped up to the process. Two types of hook-up diagram are illustrated here, one showing all the mechanical items that will be needed to complete the assembly (such as elbows, tees and unions), the other outlining in schematic form how the system operates. Normally, only one of these types will be used for a given contract. Such diagrams define the connections and as such are an essential prerequisite for installation of an instrument. The detailed version of the hook-up is useful for costing/estimating purposes, although the same information may be provided in a simpler form.

A few points about this diagram warrant further discussion. The tapping-point isolating valve is usually provided by whoever installs the main high-pressure plant pipework. It is connected to the transmitter subsystem by a small-bore line, known as the 'impulse' pipe, and the

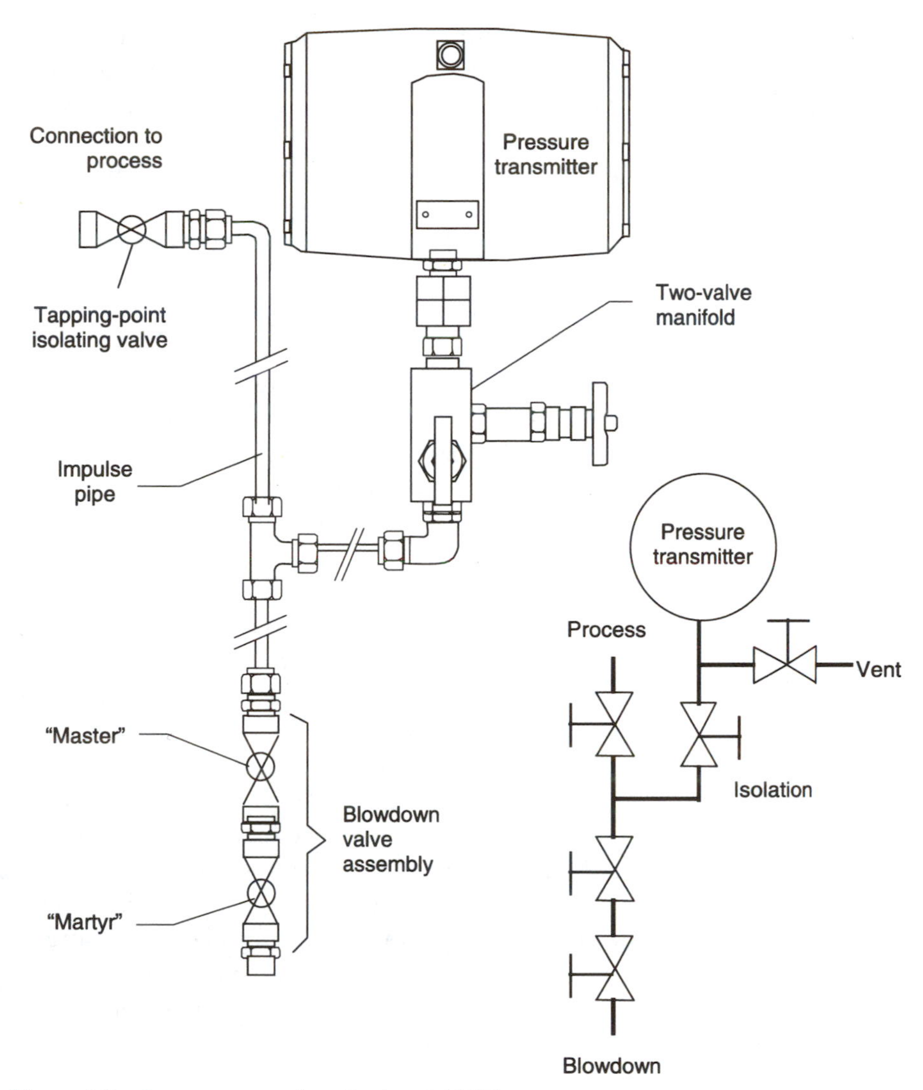

Figure 8.2 Pressure transmitter 'hook-up' (HP)

selection of the correct type and size of pipe will have a considerable bearing on the accuracy, reliability and maintainability of the installation.

The standards that are applied change from time to time and vary between countries and users. It was usual for HP pressure and differential-pressure transmitters to be connected to the process via 15 mm OD stainless-steel impulse pipes and 20 mm nominal-size valves, while drum-level transmitters were connected via 16 mm OD pipes and 32 mm nominal-size valves. Commercial pressures have nowadays led to a situation where some purchasers merely stipulate that the whole system should operate satisfactorily (sometimes within defined parameters) for a given time (say, twenty years). Little or nothing is said about matters such as maintainability of ease of access. Using the above connection sizes will lead to an installation that is workable, reliable and easy to maintain.

The sizes of the connection pipes and valves form only one part of the picture. Other matters concern the lengths, lagging, and slope (rise or fall) of the connecting pipes (which will allow them to be vented or drained), and so on.

The two-valve instrument manifold shown in Figure 8.2 is a standard subassembly which may either be integral with the transmitter or provided as a separate item as shown. It allows the transmitter to be connected, calibrated and vented before removal.

The blowdown-valve assembly enables the pipework to be flushed through to remove entrained gases, deposits etc., to a suitable drain or vessel. In this example it comprises two valves, a 'master' and 'martyr'—an arrangement that enhances long-term maintainability in a high-pressure application. If a single blowdown valve were to be provided in such an application, the differential pressures and velocities to which it would be subjected each time it is opened or closed would quickly erode the internals. The 'master and martyr' assembly operates as follows to avoid this problem. Prior to initial commissioning of the system, both valves are closed. When pressure has been applied, the master valve is opened first, followed by the martyr. When any debris or undesired gas or vapour has been ejected, the martyr is shut off first, followed by the master. When the system is to be shut off again after use, the master is opened while the martyr remains closed, and then the martyr is opened. By this means the onerous duty of opening or closing of the pressurised system to the atmosphere is always handled by the martyr, with the master merely opening or closing without changing the flow. When the martyr eventually succumbs to these harsh conditions of use it can be quickly and easily replaced while the master is closed without having to isolate the transmitter at the tapping point (a process that may necessitate shutting down the plant). Since the master valve is never subjected to harsh operational conditions it should survive for an indefinite period.

Figure 8.3 shows the actual installation of such a transmitter on a combined-cycle power plant (the isolating and blowdown valves are not visible in the picture, and only part of the impulse pipework can be seen).

Similar levels of detail must be defined for all types of transmitters and gauges. Each of these has its own peculiarities, and neglect of a simple requirement can render a vital measurement inoperative, inaccurate or unreliable. Again, reputable manufacturers will be able to provide

Figure 8.3 Installation of a pressure transmitter
Photo taken by permission of National Power plc

detailed guidance for each device. But beware of those who claim that their instruments are so simple that no such guidance is needed!

8.4.2 Actuators

In the chapters of this book dealing with various control loops, reference has been made to the controlling devices (such as valves and dampers) which translate the control system's demands into changes of flow and pressure. The modulation of these devices is the duty of an actuator, and in the next few paragraphs we shall briefly survey some of the actuator types to be found in power plant.

Pneumatic actuators are well established and cost-effective. In addition to being reliable, accurate and capable of fast response, they are simple to use and maintain. They are therefore found in a great many installations, and where they are used the control system's commands have to be processed by an electro-pneumatic positioner (discussed later) or trans-

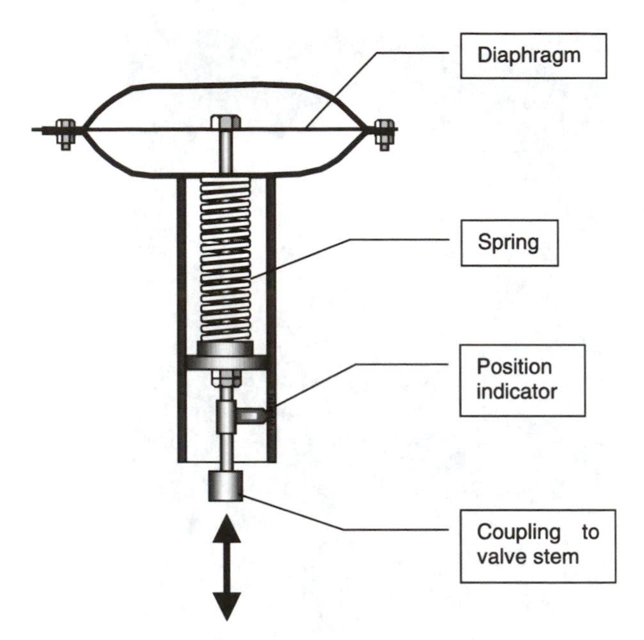

Figure 8.4 A diaphragm acuator

lated from electronic form (generally 4–20 mA) to pneumatic form (usually 0.2 to 1 bar *g*) by an electropneumatic converter (I/P converter).

The simplest pneumatic actuator consists of a diaphragm coupled to the stem of a valve (Figure 8.4). In the example shown, air pressure applied to the top of the diaphragm will cause it to be deflected and the attached stem will move downwards. When the pressure is reduced the spring will act to restore the stem position.

In many cases, the pressure available from the I/P converter will not exert sufficient force on the diaphragm to move it against the reactive force applied to the valve stem by the process fluid. Unless an extraordinarily large diaphragm is used, the 1 bar *g* output signal produced by commonly used I/P converters is unlikely to generate stem forces greater than 20 kN. Also, I/P converters are designed to feed into small volumes so that, even if they could generate enough pressure to move the valve through its full travel, they will be unable to do so quickly, since it will take time for them to build up sufficient pressure in the large volume above the diaphragm.

Two solutions are available to overcome this problem: boosters and positioners. A booster is a pneumatic relay that converts small pressures to large ones. A positioner is essentially a power controller whose function is to translate the pneumatic control signals into mechanical movement of

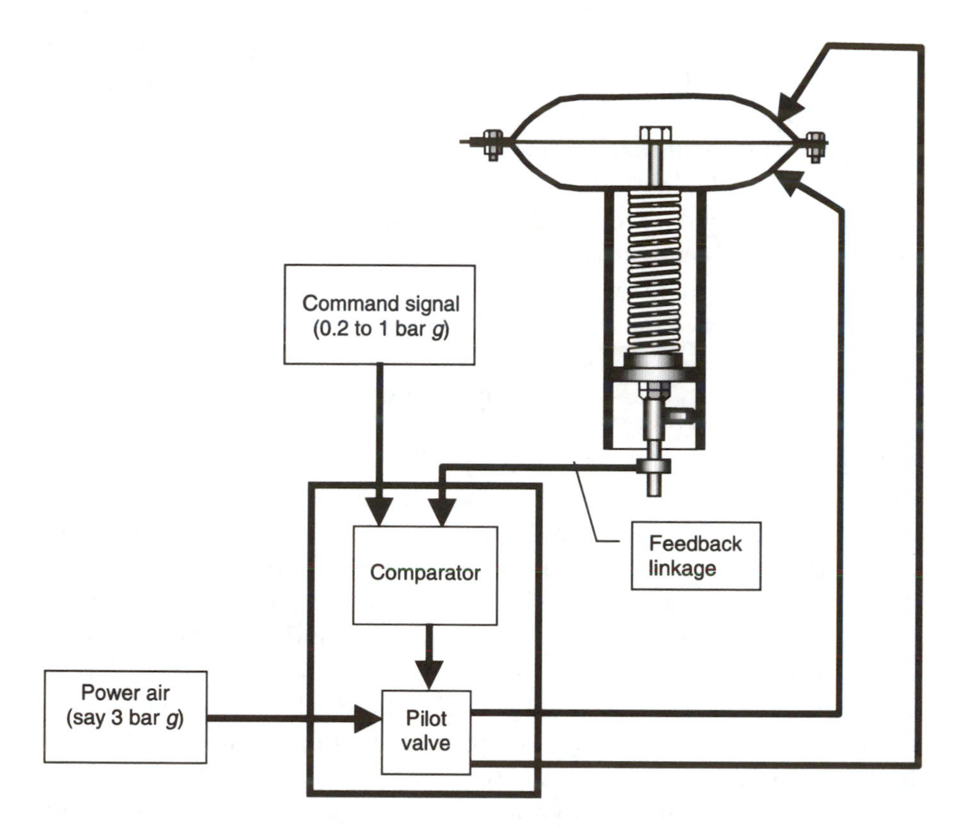

Figure 8.5 Application of a positioner to a pneumatic valve acutator

the valve stem. A positioner (Figure 8.5) applies a high-pressure air signal to one side or other of the diaphragm, adjusting the relative pressures until the stem position corresponds with the demand signal. The mechanical design of the positioner may incorporate a cam which can enable the demand/position relationship to be shaped to a linear, square-law or other nonlinear characteristic.

Although the use of a positioner is often essential, it is not always necessary. A positioner contains a feedback mechanism and difficulties can arise because of the inclusion of an additional integral term into the overall control system. This is particularly important when critical tuning is required, since the control parameters of the positioner are not adjustable.

A point which can also raise difficulties is where mechanical stops are introduced to limit the range of stem movement. In such cases it can be extremely difficult to adjust the way in which the positioner operates at the extreme limits of the valve travel. When the valve reaches the limit of movement set by the mechanical stop this will not necessarily correspond with the command signal. The positioner will therefore attempt to eliminate the apparent error by applying additional pressure to the diaphragm. If the command signal holds the valve against the mechanical stop for an extended period the volume of air on one side or other of the diaphragm will build up to the full pressure of the air supply. Once this has happened any subsequent command to move the valve off the end stop will be followed by a delay as the air pressure within the diaphragm vents off. As this venting can occur only through the pipes and valves of the system, and as these are often constricted, the vent period can easily become extended. In time-critical applications the results can be unacceptable, or even dangerous.

Where such mechanical stops are fitted it is also important that the positioner is set up to correspond with the actual range of free movement of the valve. If one stop is set at a position that is, say, 10% from the fully closed position of the valve and the other is set at 90%, the positioner should be set up so that the stem is moved to the 10% position when the command signal is at the minimum of its range (e.g. 4 mA), and to the 90% position when it is at the maximum (e.g. 20 mA). If this is not done and the positioner is ranged over the theoretical stroke of the valve stem (not between the stops), saturation will again occur as the positioner attempts to move the valve past one or other of the stops. However, even with the correct settings, it will be almost impossible to avoid the problem of saturation since, even if the difference between the position of the mechanical stop and the corresponding signal is microscopically small, over an extended time the integral-action effect of the positioner will cause the air pressure in the diaphragm to build up (or vent), with all the implications already described.

In such cases (and, in fact, wherever a fast-responding loop already exists around the valve), it may be advantageous to abandon the positioner. This aim will be easiest to achieve if sufficient force can be exerted by the signal-air pressure alone but, if this is done, the length of pneumatic pipework between the I/P converter and the diaphragm should also be minimised in order to reduce the delays that can otherwise be caused.

If the pressure available from the I/P converter is inadequate, or if it is impossible to obtain a short pipe run between the converter and the diaphragm, consideration should be given to using booster relays to provide adequate air pressure to the diaphragm without the use of a positioner.

However it is done, the point is that, in many loops, the use of a positioner should be a last resort rather than an automatically applied solution.

Figure 8.6 A proprietary I/P converter
© Watson Smith Ltd. Reproduced by permission

8.4.3 The I/P converter

As mentioned above, an I/P converter is required to convert the electronic commands from the DCS to a form that can be used by a pneumatic actuator. This functionality may be incorporated in the positioner itself (an electropneumatic positioner) or a discrete converter may be used (Figure 8.6). The type of converter shown in this diagram offers a 'fail in position' function, meaning that the output pressure is maintained at the last good value when the control signal is switched off or interrupted.

Inherent to any I/P converter is the mechanical assembly which converts the electronic signal to the pneumatic output. As this will consist of some moving mass it will inevitably be affected by vibration to a greater or lesser extent. As the valves and dampers are necessarily mounted on pipework and ducting which is mechanically coupled to moving machines such as fans or pumps, they will be prone to vibrate and this factor must be considered in deciding whether to use an electropneumatic positioner or a separate converter. The former will be exposed to the full effects of the vibration while the latter can be mounted on a nearby wall or stanchion, offering some degree of insulation from the worst of the vibration. On the other hand, it is always less costly to install one device (a single I/P positioner mounted on the valve) rather than two (an I/P converter driving a separate pneumatic positioner on the valve). Also, the distance/velocity lags that affect all pneumatic systems will be diminished if the converter is an integral part of the actuator.

8.5 Electric actuators

Although pneumatic actuators are inexpensive, reliable and fast-operating, their use necessitates the provision of compressed-air supplies. The air must be clean and dry, entailing the use of filters and driers. It is therefore attractive to consider devices that do not require such expensive ancillary plant. In addition, the compressibility of air makes it difficult to provide a 'dead-beat' response when dealing with large masses.

With the evolution of reliable solid-state position controllers for electric motors, the scope has opened up for avoiding the use of pneumatic operators by the use of electric actuators. These are self-contained, and only require an ordinary source of electric power. They can provide dead-beat response and also have the advantage of providing inherent 'fail-fix' operation since on loss of power they lock in position. On the other hand, making an electric actuator fail to the open or closed position on loss of power is not so simple.

When specifying an electric actuator it is important to state the required failure mode as well as the operating speed (time to travel from fully closed to fully open).

8.6 Hydraulic actuators

Hydraulic actuators offer another way of dispensing with air compressors and their ancillary equipment. This type of actuator is powerful, fast and accurate, and can be provided in fail-open, fail-closed or fail-fix configurations. The points to consider all centre on the nature of the hydraulic medium employed. Is it flammable or corrosive, what provision is made to guard against leakage etc. Also, if for reasons of economy a centralised hydraulic reservoir is shared between several actuators, careful consideration must be given to ensuring that no failure can disable major portions of the plant.

8.7 Cabling

The cables linking transmitters and actuators to the control system will be installed in areas where they may be exposed to impact from passing vehicles or falling objects. In addition, they may be subject to movement of the structure. For these reasons, cables should be adequately protected and well supported. It is common practice in some countries to use steel-wire armoured cable to provide protection, but if adequate mechanical support and protection is provided by other means (such as cable trays) there is an argument in favour of using cheaper, unarmoured cable, even in the most severe plant environments.

Often, a boiler is actually suspended from a steel frame to allow it to expand and contract as it heats. The movement between objects on the boiler front and a fixed reference point can be quite considerable and, unless this effect is considered, the cable can be damaged.

8.8 Electromagnetic compatibility

The high voltages, heavy currents and large magnetic fields associated with power-station equipment give rise to the risk of interference with electronic systems. It is an important requirement that the system designer recognises this fact and pays careful attention to dealing with the risk. Guidance on design and installation practice is available from several sources [1] and it is nowadays generally a mandatory requirement that systems comply with electromagnetic compatibility (emc) rules defined by the country in which the plant is to be operated [2].

In general, a system designed to be immune to interference should employ optocoupled inputs and outputs for digital signals, and its analogue inputs should include lowpass filters to provide a high level of attenuation to frequencies above say 20 Hz. Such filters provide discrimination against 50 or 60 Hz pickup from mains-operated devices and against any high-frequency disturbances that may be generated by switchgear, variable-speed devices and the like. If the roll-off characteristics of these filters are adjustable, the input channels of the system can be made to recognise legitimate variations in the measured signals (dealing with rapid pressure changes or slow-changing temperatures, for example) while effectively ignoring interference-induced signals (such as 50 Hz pickup).

8.8.1 Earth connections

Good earthing practice demands the use of a star-point connection for all screens. Figure 8.7*a* shows how current flowing through a common impedance affects other circuits connected to it. In this example, the voltage appearing at the input of a device (with reference to earth), is given by

$$V_1 = (i_1 \times z_1) + (i_3 \times z_3).$$

It must be recognised that a major fault in an electrical machine is a transient phenomenon which can result in a very large current flowing to earth in a very short time. In addition, lightning strikes on structural steelwork, cables and machine frames can cause currents of hundreds of kiloamps to flow to ground. Because of the high-frequency nature of the current in all such cases, the complex impedance of the earth connection becomes dominant (i.e. its resistance and its inductance).

If the device represented by z_1 is an analogue input channel of the plant DCS, it would normally be handling 4–20 mA signals and its input impedance would be, say, 250 Ω. The voltage across this input channel at full-scale would therefore be expected to be no more than 5 V (0.02 A × 250 V).

Now, assume that the common earth impedance is 10 Ω and that a transient fault current (i_3) of 100 A flows through it. (In practice, fault levels can be much higher even than this.) The voltage occurring at this input channel of the DCS under these conditions would therefore be

$$V_3 = (0.02 \times 250) + (100 \times 10) = 1\ 005\ V.$$

This example shows that input circuits of the DCS can be subjected to voltages several hundred or thousand times higher than expected, due to such fault currents.

Figure 8.7*b* shows how this effect can be minimised by segregating signal earth connections from the earth connections of machines and their switchgear. The currents flowing to the instrument earth will all be of the order of milliamps, and if the connection to earth has a low impedance the maximum voltage appearing at the common point will be no more than a few hundred millivolts.

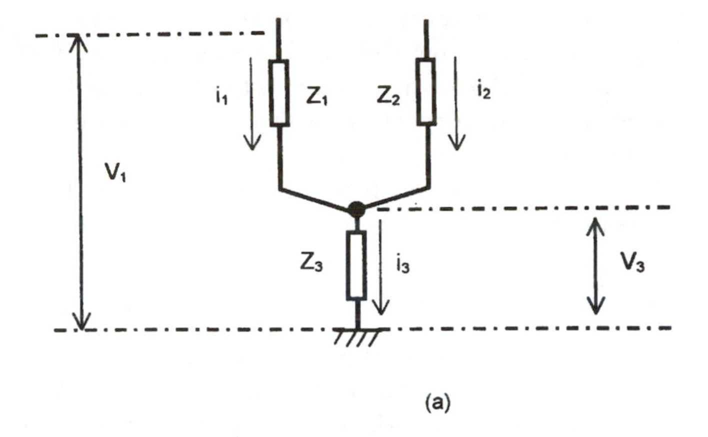

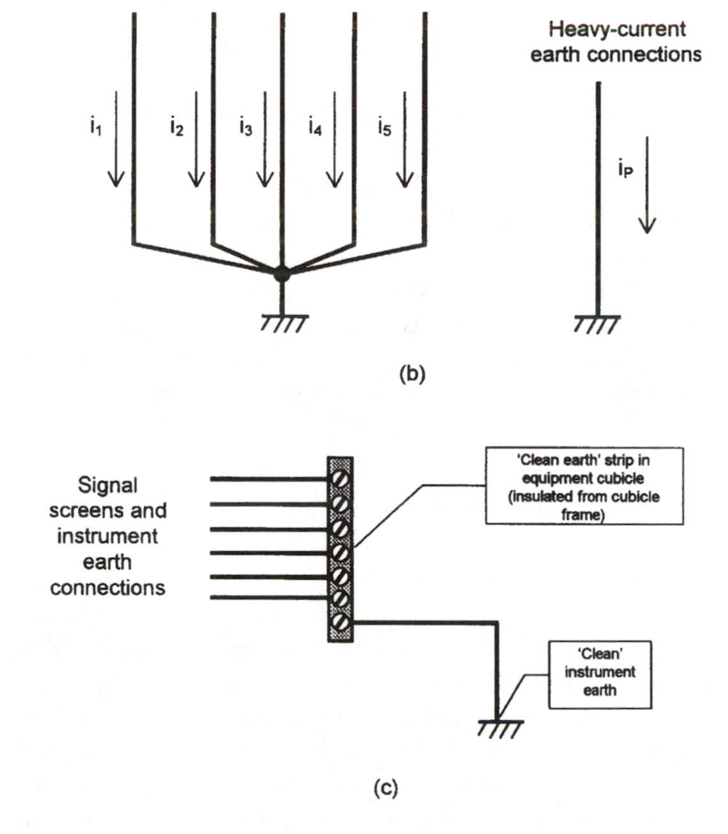

Figure 8.7 *Good earthing practice*
a Current flow through common impedance
b Segregation of instrument and electrical (safety) earth connections
c A practical arrangement for an instrument earth connection

Figure 8.7*c* shows how such segregation can be achieved in practice. A low-impedance busbar, insulated from the metalwork of the cable itself, is provided in each equipment cubicle and connected to a 'clean earth' point by a low-impedance cable. All instrument earth connections (including cable screens) within the cubicle are connected to this busbar. The metalwork of the cubicle is connected to a separate safety earth point (often through the mounting bolts).

8.8.2 Cables: armouring, screening and glands

Reference has been made above to the use of steel-wire armouring to provide mechanical protection for interconnecting cable, and it is sometimes argued that armouring provides immunity to interference. This is not totally correct. Although steel-wire armouring does provide some degree of protection against magnetic fields, its performance as an *electrostatic* screen is poor. For this reason it is essential to use cable with a braided (or foil) screen, with or without overall steel-wire armour, for all signal connections.

Figure 8.8 shows how an armoured, screened cable is used to connect between the component parts of a system. It also shows how the various conductors should be connected to earth.

Because it is very difficult to spot missed, duplicated or badly made earth connections once a cable installation has been completed, it is vital that work is properly supervised and very carefully checked during the installation of a system. In this respect, useful assistance is provided in standards such as BS 6739:1986 'Code of Practice for instrumentation in process control systems: installation, design and practice'.

8.9 Reliability of systems

Because of the large numbers of electronic components that are manufactured, and because component manufacturers keep good records of failure rates etc., it is fairly easy to obtain statistical information on reliability that will provide a good indication of the predicted reliability for any given system. In practical terms, what really matters is the length of time for which a system will be capable of remaining in operation over the course of a year or over its operational lifetime. This is governed by both the reliability of the equipment and the speed with which repairs can be effected.

For example, it would be theoretically possible to construct a very reliable system by arranging for all functions to be performed by a few very large-scale integrated (VLSI) circuits connected together without the

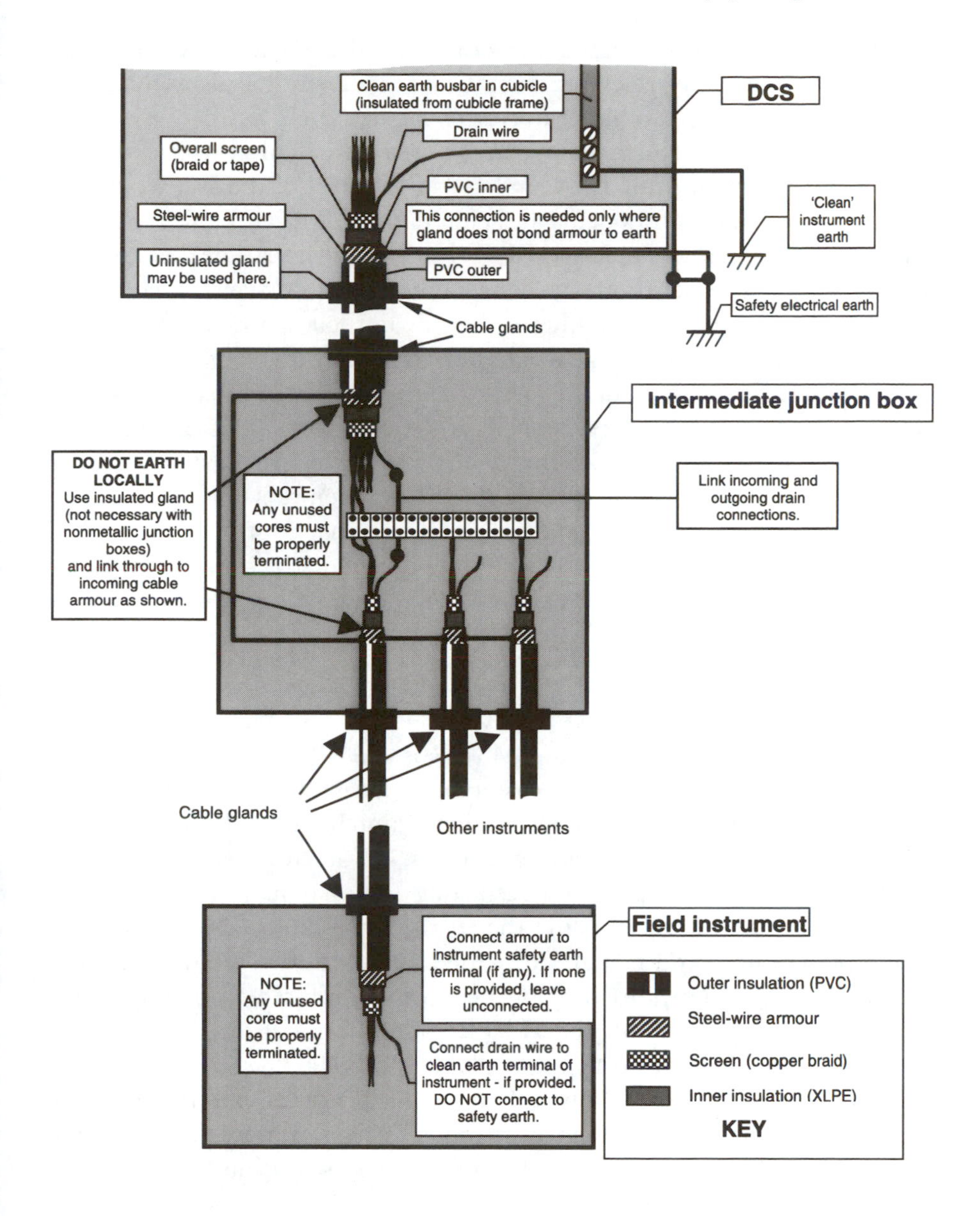

Figure 8.8 Screening and earthing of cables

use of plug-and-socket connectors. Because VLSI devices are inherently reliable and because connectors are a source of failure such a system would offer a very high level of reliability. Unfortunately, it would be very difficult to repair if it did fail.

The reliability of any electronic system can be predicted with a high level of confidence by referring to statistical data produced by manufacturers, independent test laboratories or bodies such as the defence or nuclear authorities*. Such data can be used to calculate the predicted failure rate, or mean time between failure (MTBF), of the system and by using ultra-reliable components and eliminating all less reliable devices, it should be possible to achieve MTBF rates of perhaps one failure in a million hours of operation (i.e. one fault in just a over a century of operation!). However, if a failure did occur in such a system, locating its source and repairing the fault would be extremely time-consuming. Here, another statistical calculation is used: the mean time to repair (MTTR). This figure is based on factors such as the diagnostic tools available to locate the source of a fault, the availability of spare parts, the work involved in removing the faulty component and then replacing it.

A useful way of looking at the practical aspects of reliability is to combine the two factors. This leads to another statistic, the system 'availability', which is a combination of the MTBF and MTTR

$$\text{availability} = (\text{MTBF} \times 100)/(\text{MTBF} + \text{MTTR}).$$

Using this formula shows that the availability of a system with a MTBF of 80 000 hours and an eight-hour MTTR is 99.99%. Achieving the mean time to repair of eight hours is reasonable. This is the time from the fault occurring, through the process of locating maintenance staff to carry out a repair, through the fault-finding process, to locating a replacement component, to installing it and restarting the system. If the diagnostic tools are very powerful, enabling the location of a fault to be quickly and easily pin-pointed, and if spare printed-circuit cards are mounted nearby in the system cabinets, already powered (and therefore warmed up), then it may be possible to reduce the MTTR and if this is cut to say four hours, the same system will now offer an availability of 99.995%.

When evaluating the likely reliability of a system, all three of the above factors should be examined, together, because it may be that a high level of availability is based on a less reliable configuration but an impossibly short MTTR.

At first glance an availability of 99.98% may appear to be very good, but if this is based on a four-hour MTTR it implies that the MTBF is 20 000 hours. This means that the system is likely to suffer failures on about nine occasions over an operational lifespan of 20 years. A system

*For example the Systems Reliability Service Data Bank of the United Kingdom Atomic Energy Authority, AEA Technology, Thompson House, Birchwood Technology Centre, Risley, Warrington, Cheshire WA3 6AT, UK.

with the same availability, but with a more realistic eight-hour MTTR would have a 40 000-hour MTBF, meaning that over the same lifetime, the system could be expected to fail on about four occasions.

It must be remembered that availability, MTBF and MTTR are all statistical predictions. Nothing in them will guarantee that a system will operate without fault for a defined time. (In fact the system may still go wrong on day 1, though the likelihood then is that it should not go wrong again for a very long time, although that may seem to be poor consolation at the time.)

It must also be appreciated that it is not realistically possible for these statistics to be confirmed by measurement. At best, a so-called 'reliability run' may extend for a few weeks, but this represents only a few hundred hours of operation, which is a small fraction of a typical MTBF prediction (which is usually in the order of tens of thousands of hours). A reliability run will only show up problems where the reliability is seriously deficient.

To realistically evaluate a supplier's predictions, the best that can be done is to obtain the data on which the calculations have been based and compare one system with another, while at the same time asking whether any assumptions that have been made are reasonable. Beyond that, the designer should look at what is likely to happen when the chips are down.

8.9.1 Analysing the effects of failure

In the course of designing a control loop careful thought must be applied to the effects of failure of any component. If any risk can be posed by such a failure, precautions must be taken to limit its effects. Such considerations must be applied to transmitters, process switches and actuators, as well as to the DCS itself. It will usually be necessary to have the design confirmed by some form of risk-assessment procedure such as a HAZOP (hazards and operability study) [3].

The HAZOP procedure has traditionally been applied by considering the results of failure of each and every item on the plant. One of the approaches that is adopted is for a team from each discipline to look at each item and ask a series of questions such as, for a valve: what happens if it opens, shuts or locks in position? Otherwise, the questions may be aimed at assessing the effects of more or less pressure or temperature on the device in question. The HAZOP procedure is very specialised, and the audit of the plant is usually conducted during the design stages of the project by a team of process engineers, control engineers and others, the whole being co-ordinated by a specialist organisation.

The emergence of programmable systems has raised several questions as to the validity of this type of study. For example, a traditional HAZOP may lead to the conclusion that if one valve fails to open the situation may be dealt with by the opening of another valve or by the tripping of a pump (either action being initiated by the human operator or by a safety interlock system). However, with a programmable system it may be necessary to consider the possibility of a failure in the DCS causing multiple failures to occur at the same instant, while at the same time any corrective action that the operator may wish to take, and the protective systems themselves, are disabled or seriously impaired. Such questions have recently been addressed, and the matter must be considered in the light of the new guidelines [4].

The following provides an overview of some of the safety-related matters that will need to be considered during the design procedure.

8.9.1.1 Failure modes in electronic systems

Oversimplistic approaches are sometimes adopted towards the analysis of failure, and the field of industrial control-systems design is littered with ill-considered ideas. One of the most notable of these is that 'an electronic signal will always fall to zero under a fault condition'. Worse still is the theory that this is the reason for the use of the 'live zero' (e.g. 4 mA) in signals such as the well established 4–20 mA range.

Though an electronic signal may well fall to zero under certain conditions (e.g. breakage of the connection between the transmitter and the receiver), it is as likely to rise to 20 mA (or above), to lock at an intermediate value (even though the measured parameter has changed), or to slowly drift away from the correct value. If the signal source is provided with self-diagnostic facilities (as described in the following section), the output can be configured to rapidly change to a high or low value, but this is the only condition where such an action can be assured.

As for the reason behind the use of a live zero, this has little or nothing to do with failure. The real reason is that the live zero provides a minimum current for the signal source, enabling the device to be powered from the receiver. This is known as 'two-wire transmission' since it eliminates the need to provide separate conductors to power the transmitter. Figure 8.9*a* shows a transmitter working on a 'dead-zero' range such as the old 0–10 mA standard. This requires the provision of power-supply connections to the device, so that four conductors are needed for each transmitter. With a live zero (Figure 8.9*b*) the transmitter electronics can be powered from the receiver, and only two conductors are needed.

Given that a process transmitter is capable of detecting problems within itself and warning of this, so that the DCS can take the appropriate action, what can be done about the signals transmitted by the DCS to

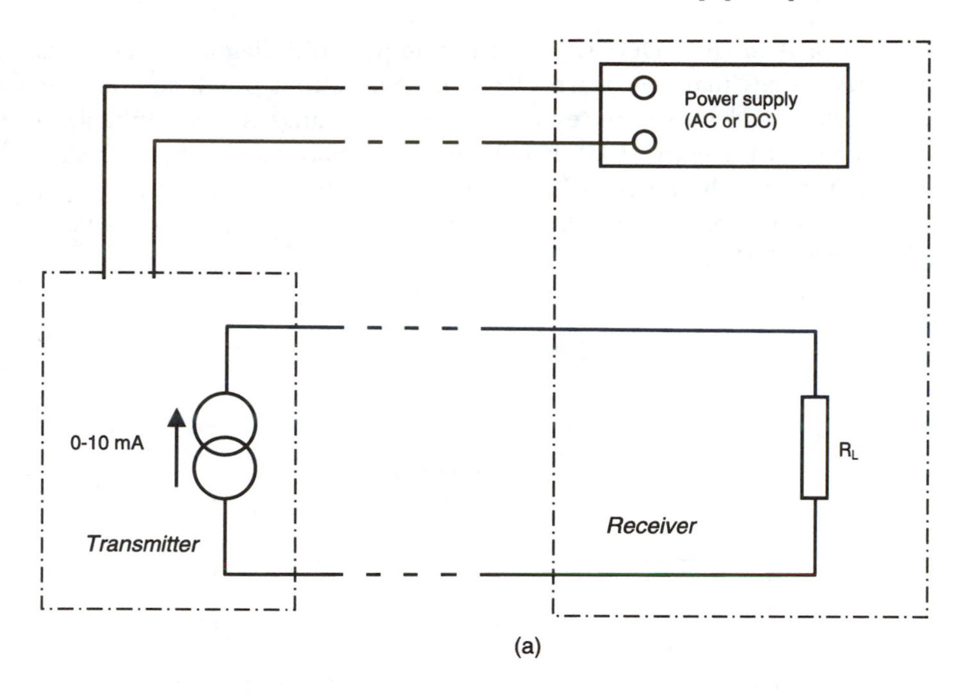

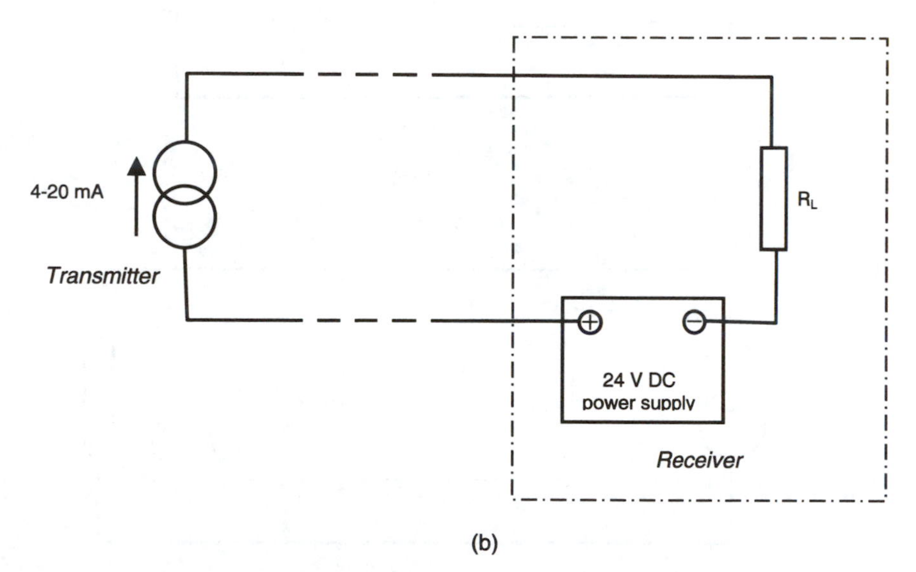

Figure 8.9 *The advantages of two-wire transmission*
a Transmission without 'live zero'
b Two-wire transmission (with 'live zero')

actuators? A modern DCS will incorporate powerful diagnostics facilities and watchdogs that will warn of incipient or actual problems. But the action that can be taken in response to these warnings is more difficult to determine. If the entire DCS is failing, then it may be necessary to shut down the plant. On the other hand, if only one output channel has failed it may be possible to override the commands or operate emergency devices that bypass the fault.

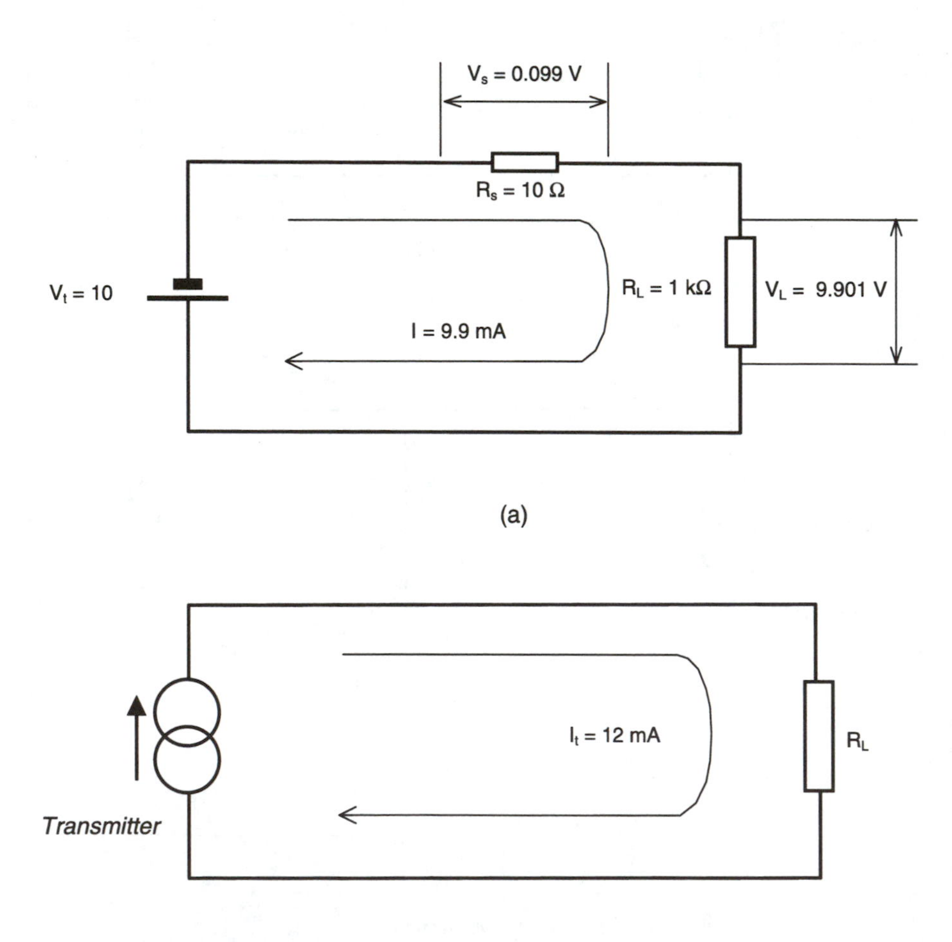

Figure 8.10 *Current vs voltage transmission*
a Voltage transmission
b Current transmission

8.9.1.2 Voltage and current transmission

The use of current rather than voltage signals for transmitters may require explanation. If a transmitter generates a voltage signal (Figure 8.10*a*), the voltage that appears at the receiver will not be the same as that at the transmitter because of the voltage loss across the series resistance of the connecting cables. In the example shown, the receiver input resistance is 1 000 Ω, and the total cable resistance is 10 Ω, so the current that circulates when the transmitted voltage is 10 V is 10/1 010 A, or 9.9 mA. The voltage drop across the 10 Ω series resistance causes a voltage drop of 99 mV, so the voltage appearing at the receiver is 9.901 V. In this example the signal appearing at the receiver is roughly 10% below the transmitter output.

The situation is made worse because the cable resistance is not fixed. It will vary with temperature and will also be affected by factors such as the contact resistance of the terminals and any switches in the circuit.

The effects can be minimised by using a receiver with a high input resistance, but however high the resistance is made, some small current flow will always exist, and therefore some errors will be introduced.

With a current system (Figure 8.10*b*), the current sent out by the transmitter is by definition identical to the current flowing through the receiver terminals. The only possible source of error would be the formation of a parallel path across the receiver, which would divert some of the current flow, but this is very easy to avoid.

Figure 8.9 compares the circuit of a voltage-based signalling system with a current-based equivalent. It will be seen that the voltage-based system requires additional connecting wires for the power supply, implying the need for four wires, whereas a current-based system requires only two. (In figure 8.9*b* the power supply is connected between the negative end of the load and the positive terminal of the transmitter. However, the power unit could equally be moved to the other side of the circuit, which is preferable with some types of circuit because of the grounding arrangements of the various devices. Each system must be carefully examined in order to verify the optimum method of connection.)

A further benefit of a current transmission signal is that its sensitivity to interference is less than that of a voltage system. In a current system, if interference causes the current to increase, the transmitter will quickly self-regulate and restore the current to the correct value. With a voltage system, a voltage induced by interference and that generated by the transmitter will be summed at the receiver input terminals. The transmitter will not be aware of this, and so cannot correct for it.

8.9.1.3 Transmitters

Modern transmitters are frequently able to run self-diagnostics routines that provide a high level of confidence that a failure will be detected. In general these routines are arranged to drive the transmitted signal to one extreme or other (usually outside the normal operating range of the instrument). The direction in which the signal is driven must be defined when the instrument is first specified.

Thus a 4–20 mA transmitter can be specified to drive the signal to under 4 mA or above 20 mA when a fault is detected. The DCS must then be configured to raise an alarm and take the necessary action to protect the plant when such an event is detected. Depending on the hazards which may arise, the required action may be to 'freeze' the relevant system output command at the value it had held prior to the fault being detected, or it may be to open or close an isolating valve, or it may be to operate a separate emergency shut-down system.

The fault-detection process will drive the signal to the selected point virtually instantaneously, and if the DCS is configured to respond quickly to such an event the necessary action will usually be taken before any hazard can arise.

If the nature of the process or its instrumentation is such that it is not possible to provide adequate protection by the means outlined above, some other form of protection will be needed. This may take the form of a measurement and control system that monitors the process by an entirely separate set of instruments and takes action if a dangerous discrepancy arises between the two sets of measurements, or it may be a sophisticated two-out-of-three (2oo3) voting system. Various guidelines are available for making such decisions [3,5], and these should be consulted to justify the cost of providing adequate levels of supervision and back-up. It is unlikely that any authority would be able to supervise each step in the design, construction and installation of a power-station DCS. But once the plant has commenced operation, if an incident does occur and results in damage to the plant and/or injury to personnel, the entire design and construction process will be very closely examined. The consequences will be severe for all involved if at that stage it is not possible to demonstrate compliance with the various guidelines and standards.

8.10 Summary

Armed with an understanding of the plant and of its control and instrumentation systems and equipment, we can now move on to look at how the operational requirements are defined on a project, and some aspects of how equipment on a plant is identified.

8.11 References

1 BULL, J. H: 'Code of practice for the avoidance of electrical interference in electronic instrumentation systems'. ERA 75-31, ERA Technology Ltd., Leatherhead, Surrey, UK
2 'Electromagnetic compatibility for industrial process measurement and control equipment'. IEC Publication 801-1. International Electro-technical Commission, Geneva, 1987
3 'A guide to hazard and operability studies' Chemical Industry Safety and Health Council of the Chemical Industries Association, London, 1992.
4 MINISTRY OF DEFENCE DIRECTORATE OF STANDARDIZATION: Draft Standard 00-58: a guide for HAZOP studies on systems which include a programmable electronic system'. (Ministry of Defence, Glasgow, UK, 1995).
5 HEALTH AND SAFETY EXECUTIVE: 'Programmable electronic systems in safety-related applications'. (ISBN 0 11 883913 6 ISBN 0 11 883906 3) (HMSO, London, 1987)

Chapter 9

Requirements definition and equipment nomenclature

9.1 Overview

The provision of a C&I system for a power station is a complex matter which requires careful and comprehensive administration. The task is demanding: the design of the equipment must be correct, systems must be designed on time, equipment has to be carefully specified and purchased, and everything has to be delivered to site, installed and commissioned to a tight programme which interweaves the C&I system with the many other activities that will inevitably be taking place on site at the same time. When complete the system should be fully supported by comprehensive documentation which enables maintenance staff and users to deal with it.

The actual process of designing the C&I systems forms only one part of the many activities that go together in the task of engineering a complete contract. Although some of the other operations may seem mundane and trivial, they are really anything but that. They are as essential to the contract as the technical design work.

9.2 Defining the requirements

The outcome of a major engineering project such as the design, installation and commissioning of the C&I systems of a power station will to a large extent depend on specifying the requirements at the outset so that what is provided fits within the budget and is fit for the intended purpose. Equally exact documentation is required at each stage as the system metamorphosises from the original concept to the final, functional form.

This procedure requires a considerable amount of definition, and the following outline lists the documents that might be required over the lifetime of a typical project, listed in the order in which they may be expected to be generated. This does not pretend to be an absolute definition that must be rigorously followed on every installation. It is a practical system that has produced good results when followed on several projects. Other documentation systems offered by a system vendor may be perfectly acceptable, provided that the same degree of definition is achieved at each stage.

9.2.1 The functional specification

A system can be defined in several different ways, but an essential requirement is the definition of what the system should do *in relation to the process* it is monitoring and/or controlling. This requirement is met by the Functional Specification (FS). This is a process-related definition of the functions that the system will be required to perform. It does not provide detailed descriptions of the system hardware and software, such as response times, power supplies, environment etc., except where these are critical to meeting the functional objectives of the installation. Because it defines the requirements, the FS should be one of the documents against which the vendor is invited to submit a commercial bid for a project.

A typical FS will describe the plant as a whole, and then discuss the control loops with the required accuracy, response times and dynamic range of each loop.

9.2.2 The technical specification

Having said *what* the system should do, it is then appropriate to define, in some detail, *how* the functions should be achieved. This purpose is served by a Technical Specification (TS) which describes the system configuration in terms of the electronics and computer technologies to be employed. Because it defines the technology and facilities required, this document should be requested from the vendor when he submits a proposal for executing a project.

The TS should include the following definitions:

- the size, type and number of operator displays to be provided;
- the overall configuration (heirarchical, distributed etc.);
- method of programming (block-structured, Boolean, ladder-logic etc.);
- the physical environment in which the equipment will be expected to operate (defined in terms of temperature, humidity, vibration, shock etc. as well as dust levels to be encountered, hazardous-area requirements etc.);

- the performance required (speed, response time, availability, memory etc.);
- power supplies available (including voltage and frequency excursions);
- safety requirements (backup, redundancy, fault tolerance etc.).

The TS will also lay down the requirements for testing, such as:

- Factory-acceptance: where the system is set up in the supplier's premises, connected to a simulator, or to switches and signal sources for inputs and to meters and lamps (or LEDs) to show outputs, and then put through a series of routines to show that it performs as required.
- Site-acceptance tests: performed after the system has been installed and commissioned on site, when it is subjected to changes in desired-value settings, simulated equipment trips and so on, to prove that it reacts correctly and in good time to such events.
- Reliability run: where the equipment is left in full control of the process, to demonstrate that is is capable of operating correctly for a defined period, with no malfunctions. The document should state the duration of the reliability run, the conditions under which this test will be expected to operate and what should happen if a failure occurs (e.g. start again and repeat the test).

Any commercial requirements relating to guarantees, performance bonds etc., should be defined separately, although these will interrelate with the TS and should therefore be referred to.

Note that the various acceptance test procedures will at this stage be defined only in very general terms. A full definition is provided by test specifications (as described later).

9.2.2.1 Making provision for site tests

It is usual to retain a sum (typically 5% of the overall contract value) which is paid only when the vendor has proved that the system is capable of performing satisfactorily. However, it should be recognised that a control-system supplier can demonstrate that his equipment and systems are capable of functioning as required only if the plant is made available to him for testing the system's performance on the operating plant. It is unreasonable to retain what may in fact be a substantial sum of money without giving the supplier a reasonable opportunity to prove that his equipment is as accurate, fast-responding and reliable as required. Yet, it is quite common for the plant owner to procrastinate over performing such tests. The reason for this is that a power plant represents a major investment, and starting the recovery of that investment must naturally commence as quickly as possible. As soon as the plant has been completed to the point where it can start earning its keep, strong commercial pressure comes into play, requiring the plant to operate at maximum output. The pressure will be to start earning revenue as quickly as possible and for as long as possible, to maximise earnings. Reducing the output of the plant for the purpose of carrying out tests is therefore unpalatable.

It is in everybody's interests that this dilemma is recognised and a suitable form of words developed to cover the situation, yet this is not often done. One solution is to relate the financial retention to the test programme, defining when the tests will be performed and adding words to the effect that if the tests are not carried out within a defined time after commissioning has been completed, performance retentions will be released provided that the delay is for commercial reasons outside the control of the C&I supplier.

9.3 The KKS equipment identification system

Each item of equipment on a power-plant site has to be identified by a method which enables it to be uniquely defined, specified, purchased, installed, commissioned and maintained, and this requires some logical way of numbering equipment.

Although several systems of nomenclature can be identified, two methods are in widespread use: one American the other European. The latter is knows as the KKS (Kraftwerk Kennzeichen system translated as power station designation system), which was developed by a consortium of large manufacturers and users under the auspices of the German VGB Technical Committee on technical classification systems [1].

KKS nomenclature is extremely comprehensive, and once it has been understood the system provides a very useful method of identifying any piece of equipment and its operational role in a plant. The system has its weaknesses, but it is so widely used, both in Europe and wherever European influences are felt, that it is very important for the control system designer to have at least a rudimentary understanding of its operation.

KKS defines everything on a plant, from the smallest electronic component to the largest turbo-alternator and even covers the buildings that contain it all.

Unfortunately much of the English-language explanatory documentation that is available has been translated from the original German in such a way that the meaning is sometimes obscure, and fine levels of classification have often become blurred. (For example, the distinction between 'ancilliary systems' and 'auxiliary systems' may not be immediately obvious.) The system also uses alphabetic characters which do not readily relate to English-language names of the equipment or function. For example, valves are prefixed AA, and the designation letter C is used to identify the purpose of an instrument. This can add to the confusion, since many engineers relate C to a controller. Further confusion can be caused because KKS uses the letters DP to indicate a pressure instrument used in conjunction with a control system or DCS, when DP is usually taken as referring to differential pressure.

Those who are used to the American system of nomenclature may find it confusing that the KKS codes for all equipment are determined by the functional area of the plant *to which the equipment relates*. For example, in a steam-temperature loop where the measured variable is the temperature of the steam at the final superheater outlet, the transmitter will be have a KKS code beginning with LBA, but the spray-water control valve, as part of the feed-water system, will be given a code beginning with LAA. This is very different from the American system, which allocates a conzsistent area code *to the whole loop*, and this can be confusing. It is particularly important that this distinction should be clearly understood by plant personnel, since with the KKS nomenclature the tag number allocated to the valve merely indicates that it is handling water, and gives no indication that the valve is controlling the steam temperature. Critics of the KKS system use this as evidence against the system, stating that when the completed installation is in place, the nature of the medium being handled by the valve is fairly obvious (because of the size, nature or colour-coding of the pipework), whereas the effect of the valve is not. With the older system, the fact that it is a spray-water valve is apparent from the tag number.

The differences are illustrated in Figure 9.1, which shows how the two systems are applied to the same type of control loop. It will be appreciated that, on the actual plant, the tag number allocated to this valve will merely identify it as being part of the feed-water system, and it will not be apparent that its controlling effect is on the steam temperature.

9.3.1 The importance of agreement and co-ordination

The KKS system allows different users to adopt different approaches, stressing that all parties involved on the project should agree on the selected path to be followed. Although this does show a degree of flexibility, on a given project it becomes confusing to those who were not party to the decisions that were originally made. If only for this reason, it is important that an agreement on the numbering methodology is reached as soon as work commences on a project. It is equally important that a single individ-

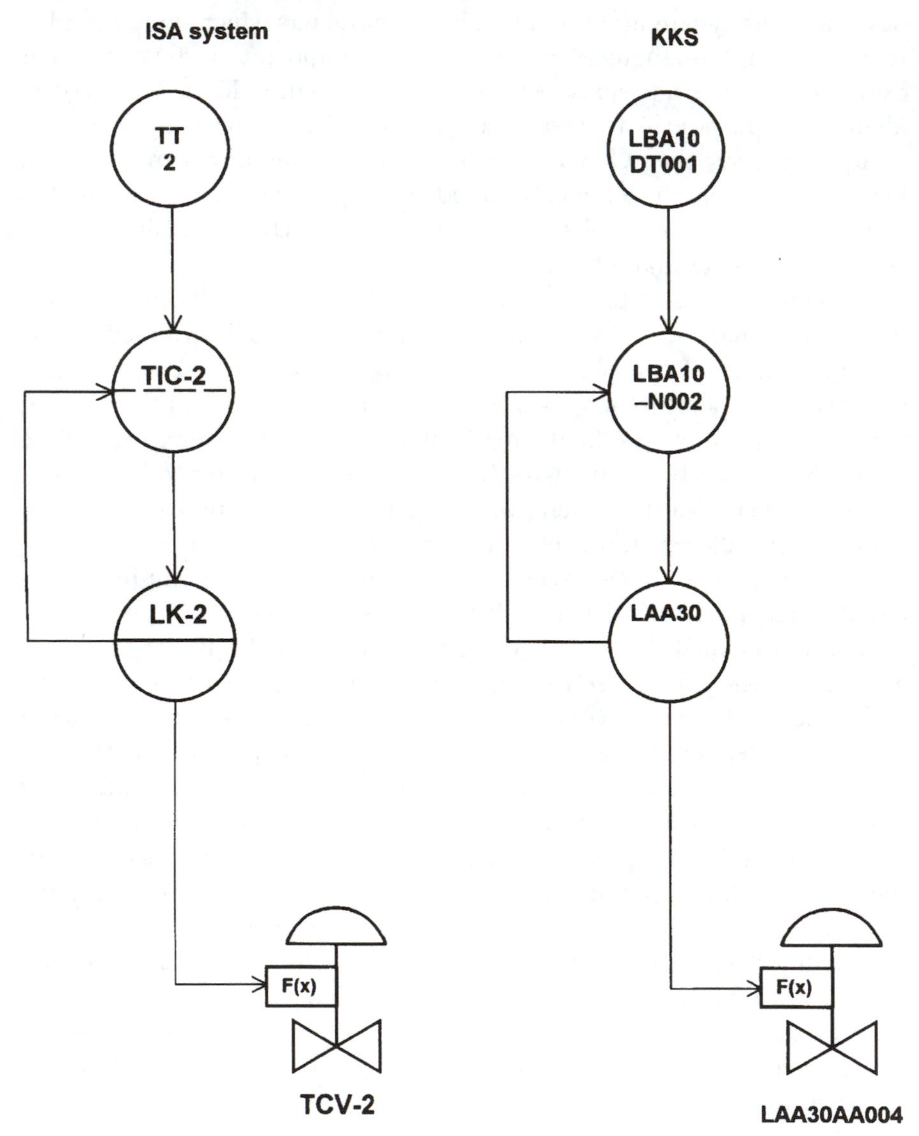

Figure 9.1 Comparison of KKS and ISA numbering systems applied to a steam-temperature control loop

ual is appointed to control and co-ordinate the allocation of numbers. This will avoid duplication and different symbols being applied to instruments that serve the same sort of function.

It is also important to understand that the KKS code for an item is almost invariably dictated by the functional area of the plant on which it is used (an exception is the DCS itself, since this generally serves almost all areas of the plant). Therefore, before a number can be applied to, say, a temperature transmitter, it is necessary to know the KKS code for the pipe to which it is fitted. This dependence renews the pressure for the plant piping and instrumentation diagrams (P&IDs) to be defined at an early stage in a project and for them to be as accurate and complete as possible before work begins on specifying the instrumentation and systems. This point cannot be overemphasised. Problems will inevitably arise as soon as any attempt is made to allocate KKS numbers to instrumentation before the P&ID stage has been completed.

It must be recognised that the KKS system offers a comprehensive method of housekeeping which can greatly enhance the task of allocating nomenclature to all equipment on a plant. This is a routine, but surprisingly costly, facet of the administration of a project. Properly used, KKS can yield considerable economies and efficiency: poorly administrated it becomes a nightmare.

9.3.2 Review of KKS

The following is an attempt to explain how KKS identifies the components that go to make up the C&I systems (and *only* the C&I systems) of a power plant. The structure of this book does not allow for a comprehensive guide to be provided, but the following brief outline should at least enable the reader to grasp the principles.

9.3.2.1 The levels of coding

The number allocated by the KKS system to a piece of equipment is broken down into a number of sections or 'levels' (see Figures 9.2 and 9.3) and within each of these levels is a field or set of fields, each field being occupied by a letter or number. Each letter or number is allocated a field identification: G for the plant, F for the function, A for the equipment and B for the component.

- The Level 0 code is used where there are, for example, two power plants on the same site. These are usually designated as 'A' and 'B', but the system allows numeric characters to be used if desired. If only one plant exists on the site the first character is omitted altogether.
- The first digit of the Level 1 code identifies the boiler/turbine unit on which the relevant piece of equipment is fitted. This is always a numeric character and is usually 1 for Boiler 1 etc. Where the equipment is common to all areas of the plant (e.g. a cooling tower) the number 0 is allocated to it.

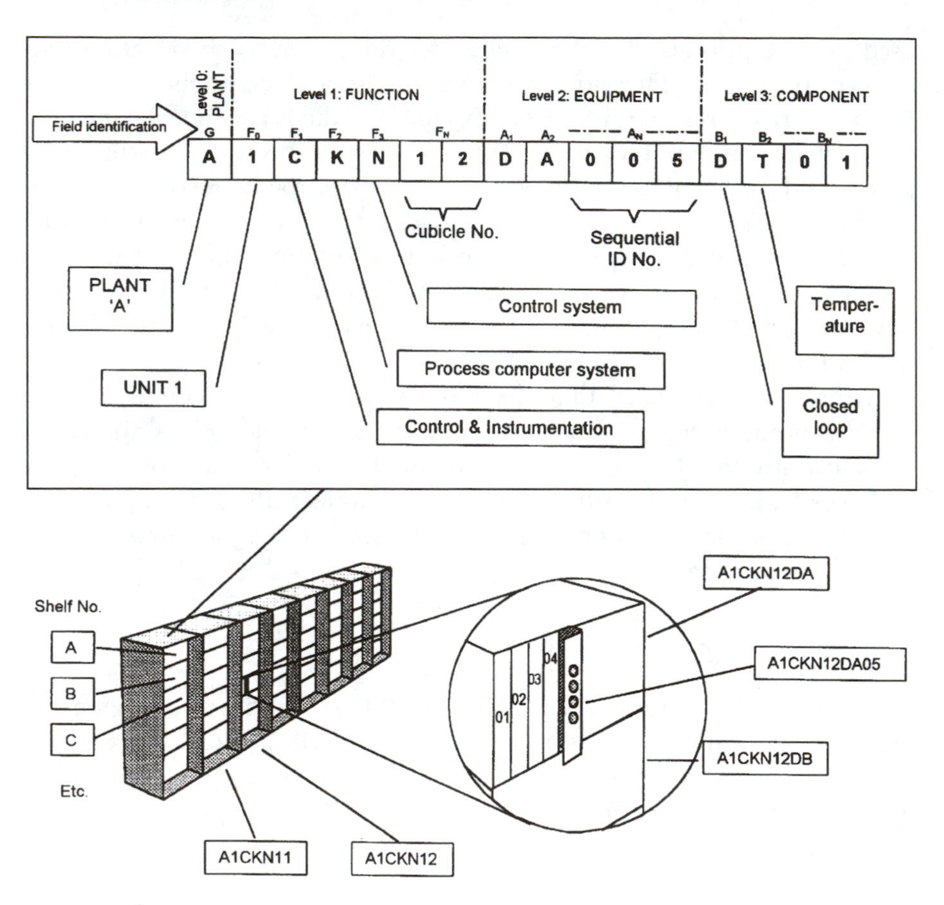

Figure 9.2 Application of the KKS numbering system to DCS equipment

- The remaining part of the Level 1 code defines the function of the equipment (e.g. a feed-water pump system, comprising the pump itself, its motor and all equipment associated with it, such as leak-off valve, starter etc.), together with a unique sequential number for the system (e.g. pump system no 2). This group of digits (the first three alphabetic, the next two numeric) is combined with the unit-identifying prefix, into what is called the 'function code'.
- The Level 2 code (which may comprise five or more digits) defines the particular piece of equipment (such as a pressure transmitter) and its unique sequential number (e.g. 101).
- The Level 3 (component or signal) code identifies the component itself, and in the case of a device that generates some form of electrical signal, it also defines the nature of that signal.

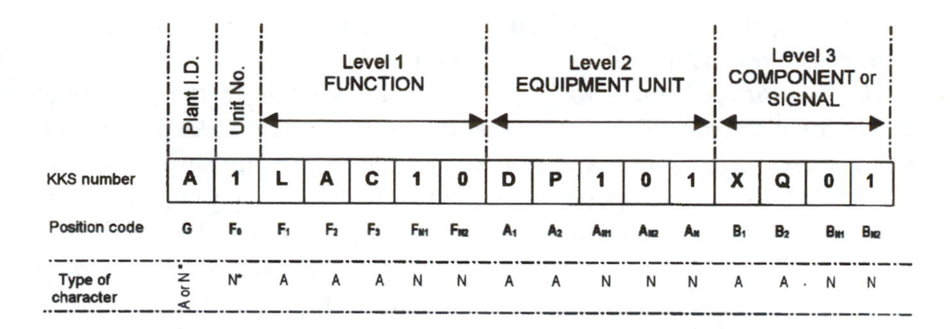

Breakdown of the code

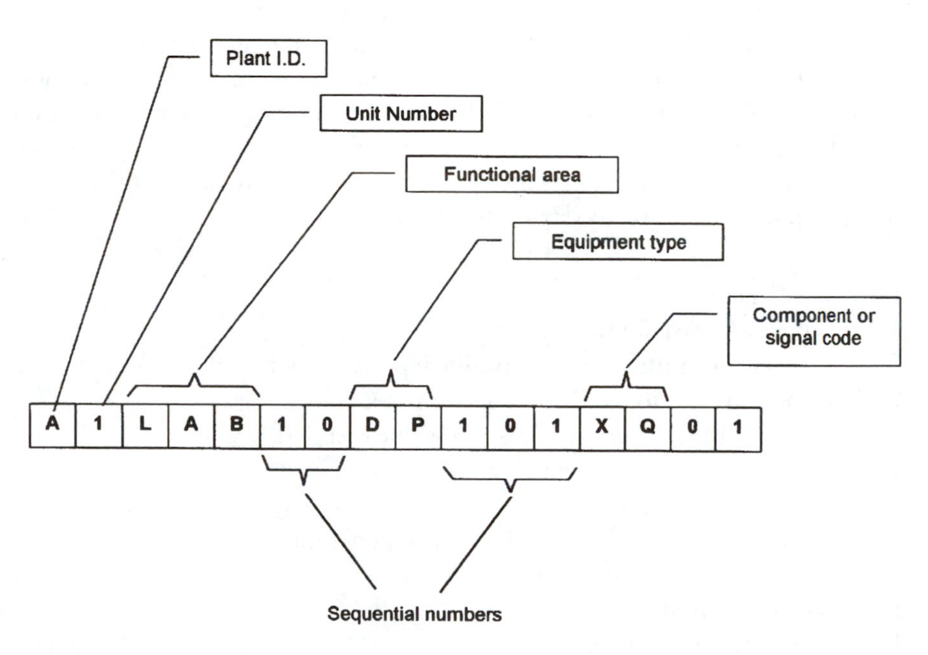

Figure 9.3 Overall KKS numbering philosophy

It is important to understand that if there are, for example, 200 pressure gauges on a plant, the sequential numbers for these do not start at 001 and continue through to 200. The numbers relate to the functional area of the plant on which they are used. Therefore, they start afresh each time the preceding code changes. For example the first pressure gauge on the HP steam piping system of a plant (where the piping system is numbered LBA10) will be allocated a sequential number LBA10CP001, the next will be LBA10CP002, and so on. The numbering starts again on another system: the first pressure gauge on the hot reheat steam piping system (numbered LBB10) being allocated the sequential number LBB10CP001, the next being LBB10CP002 and so on.

The structure and application of this coding system is illustrated in Figures 9.2 and 9.3.

9.3.2.2 Explanation of the main coding principles

There are several hundred categories within the KKS codes but the following is a brief summary which will provide some assistance in dealing with boiler C&I systems. Where it is felt that the wording is ambiguous, some examples have been provided. However, it should be remembered that these interpretations are not universally implemented, and the importance of project-specific agreement and co-ordination (as mentioned in Section 9.3.1) cannot be too strongly emphasised.

Level 1: Function
A = grid and distribution systems
B = power transmission and auxiliary power supply
C, D = instrumentation and control equipment
E = conventional fuels supply and residue disposal
F = handling of nuclear equipment
G = water supply and disposal
H = conventional (i.e. nonnuclear) heat generation
J = nuclear heat generation
K = reactor auxiliary systems
L = steam, water and gas cycles
M = main machine sets
N = process energy (e.g. district heating)
P = cooling-water systems
Q = auxiliary systems (e.g. air compressors)
R = gas generation and treatment
S = ancillary systems (e.g. heating and ventilation)
U = structures

W = renewable energy plants
X = heavy machinery
Z = workshop and office equipment

Level 2: Equipment unit
A = mechanical equipment (e.g. valves, dampers, fans, including actuators)
B = mechanical equipment (e.g. storage tanks)
C = direct measuring circuits (e.g. local indicators)
D = closed-loop control circuits
E = analogue and digital signal conditioning
F = indirect measuring circuits (e.g. sensors feeding remote indicators)
G = electrical equipment (e.g. cubicles, junction boxes, generators, inverters, batteries, lightning-protection system)
H = subassemblies of main and heavy machinery (e.g. bearings)
J = nuclear assemblies (e.g. absorbers, moderators, shielding equipment)

9.3.3 An example of how the codes are used

Figure 9.4 shows how the KKS codes may be applied to a part of the feed-water pumping system on a plant. In this example, the pump relates to Boiler 1 and the pressure indicator at its discharge is therefore numbered 1LAB25CP001, the C being the Level 2 code for a direct-measuring device (C) indicating pressure (P). The feed-flow transmitter forms part of the three-element feed-water control system and its Level 2 code therefore uses D for a closed-loop control circuit, and F for flow.

When it comes to transmitters whose signals do not form part of a closed-loop control system, different interpretations are applied by various users to the Level 2 codes. In some cases the classification letter D is applied to any transmitter which feeds the DCS, irrespective of whether or not the signal is used in a closed-loop control system. Other users extend the meaning of F (indirect measuring circuits) to include this type of measurement. In Figure 9.4, the former of these interpretations is used, and the temperature transmitter that is used to provide a signal for the operator display, is therefore allocated the letter D as the first character of its Level 2 code (the second character is T, for temperature).

This point illustrates the importance of co-ordination and agreement between all parties mentioned at the beginning of Section 9.3.1.

When the Level 3 component code is considered, the KKS classification serves two functions. One identifies the component itself, for example QT01 for a thermowell, while the other designates the signal. The signal codes are broken down into classifications which identify whether they are

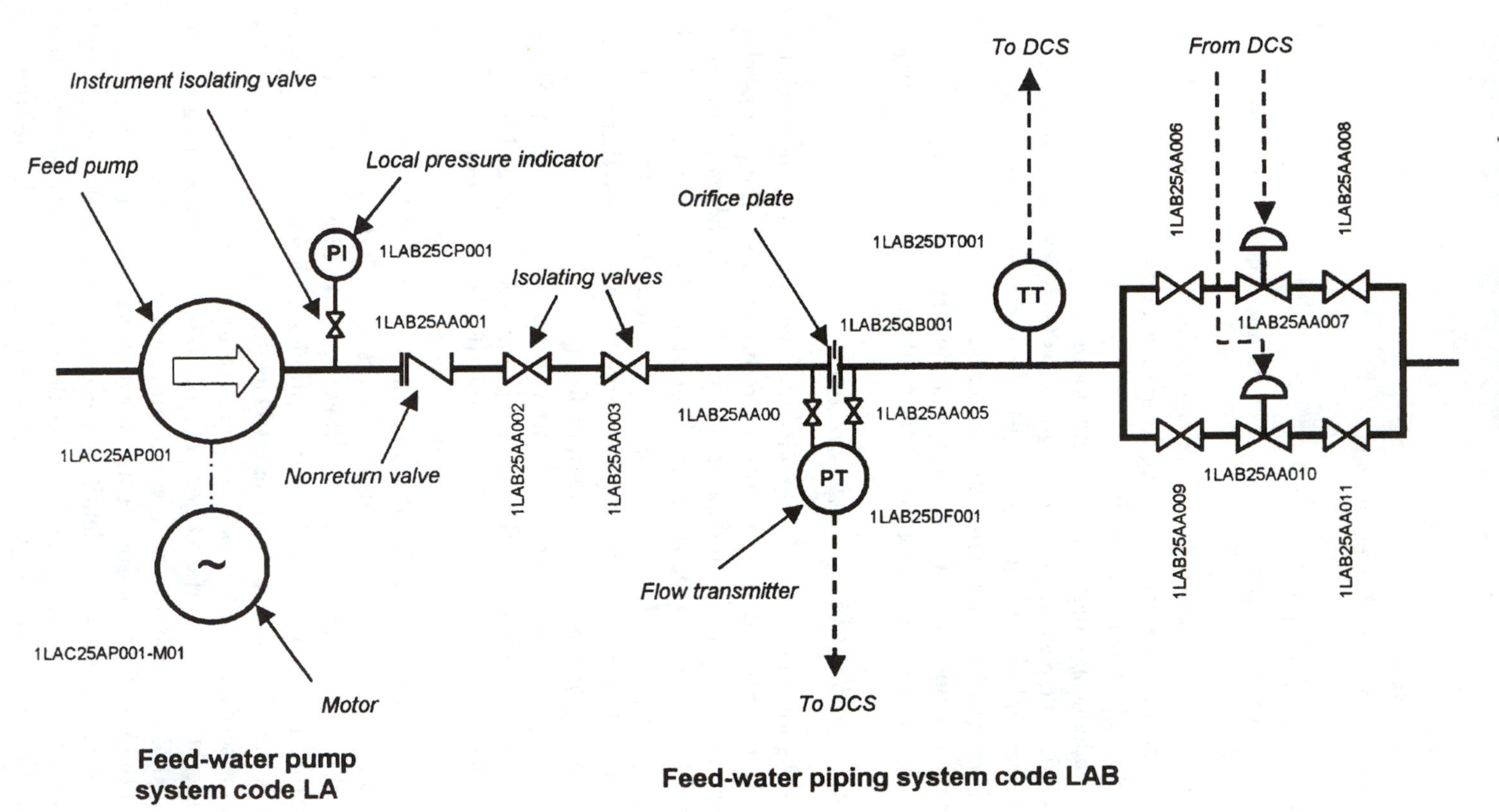

Figure 9.4 KKS coding applied to a feed-water pump system

Table 9.1 Some KKS Level 3 codes for DCS inputs and outputs

Function	Field identification B_1	B_2	B_N	Class of signal	Signal
		B	11	Binary–MCC	Alarm
			91	Binary–MCC	Other
			31	Process switch	LOW limit switch 1
			32	Process switch	LOW limit switch 2
Input to DCS	X	G	33	Process switch	LOW limit switch 3
			41	Process switch	HIGH limit switch 1
			42	Process switch	HIGH limit switch 2
			43	Process switch	HIGH limit switch 3
			91	Process switch	Other
			01	Analogue input	Resistance thermometer element
		Q	11	Analogue input	Transmitter
			91	Analogue input	Other
		B	01–05	Binary–MCC	Various START commands
			09	Binary–MCC	STOP
		G	01	Binary output	Solenoid valve
Output from DCS	Y		91	Binary output	
			11	Analogue output	Speed control
		Q	21	Analogue output	Position control
			91	Analogue output	Other

transmitted to or from the DCS. They can also show the function of the signal. Table 9.1 gives some examples to show how this is achieved.

In general, an X in the B_1 field indicates an input to the DCS and Y shows an output from it. In the B_2 field, B indicates a binary signal associated with motor controllers; G is for binary signals such as limit switches, pressure switches, solenoid valves and so on; Q is for analogue signals.

By careful use of these identifications it is possible to economise on the work of defining the functions of various signals, since each function will fall into a defined group, for which the actions and connections are already defined.

9.4 Summary

This almost completes this study of boiler C&I systems. However, updating and refurbishment needs to be discussed. This is a subject that has become important as the pace of electronics development has accelerated. As there is little sign of this trend levelling out, we must look at its implications for power-plant management, users and maintenance staff.

9.5 References

1 DIN 40719 'Circuit diagrams, circuit representations, definitions, classification (diagrams, charts, tables, item designations)—alphabetical classification examples, 1978

Chapter 10

Upgrading and refurbishment systems

Even the newest and most advanced of control systems eventually becomes dated. Up to about the middle of the twentieth century, C&I systems were mainly pneumatic, the pace of development was slow and it was generally possible to keep a system in good functional order, however old it may have been, simply by repairing the equipment or, if the worst came to the worst, buying replacement components from the original manufacturer. (In those heady days manufacturers could be relied on to remain in business for more than a few years, and to be able and willing to provide replacement components at reasonable cost for as long as the plant itself remained in operation!)

This happy situation persisted for a while after the first (analogue) electronic systems arrived, but the advent of computer-based systems eventually brought about a sea-change in the maintenance of equipment and systems.

For the hapless end-user, it often seemed that as soon as a new system was installed it had become obsolete. To make matters worse, the replacement was not only better, faster and cheaper but it was so widely used that it quickly began to be difficult to find maintenance staff who were able to understand and work with the 'old fashioned' equipment that had so recently been the owner's pride and joy. Worse still, as time went by the replacement parts became steadily more expensive and difficult to obtain.

The ultimate disaster would be when the manufacturer of the system was taken over or went out of business entirely. In the takeover situation the new proprietor would sometimes try and persuade the end user to replace the old system with one from his own stable. If the pressures were resisted, maintenance of the old system could become very expensive indeed!

It's not altogether a desirable situation, but it's what we have, so in this chapter we shall look at the problem and provide some advice on how to live with it. We start with a quick look at the reasons behind the situation, the intention being to show that, rather than being due to manufacturers trying to bolster their sales revenues, there are legitimate reasons for the constant evolution of systems and technologies.

10.1 The reasons behind the changes

The development of a computer-based system is a very complex, time-consuming and expensive business. The need for a new system is usually brought about by a competitor offering something better and cheaper (or at least apparently more cost-effective), and when that happens sales of the present generation begin to collapse. Clearly, if this situation is allowed to persist it will ultimately lead to the demise of the company, so a project has to be launched to produce the next-generation system.

At this point the systems manufacturer must look at the latest offerings from the manufacturers of the microprocessors—the components that are at the heart of all modern systems. For such companies, industrial process-control systems are not among the most valuable applications, because sales into this market are small in volume yet face the most stringent requirements for performance, reliability and safety.

The manufacturer of a domestic washing machine will buy very many more devices and will be far less demanding with respect to the performance and reliability requirements.

Reliability and consequential loss are major issues in the process-control market. A washing machine that lasts for five years seems to be almost the exception these days, whereas power stations have to operate over a lifetime of 20 years or more. A failed device on a washing machine may in the long term result in compensation claims, but these are likely to be small in comparison with those resulting from an incident in a power station. When the plant is a nuclear power station, the costs can become enormous, and the implications of failure are daunting. (After the incident at Unit 2 of the Three Mile Island nuclear power plant in the USA on 28 March 1979, one minicomputer manufacturer placed a complete and worldwide embargo on the use of their products on any nuclear plant. This presented something of a problem to the many C&I systems vendors who were already completing the manufacture of systems based on these products, for installation on nuclear power stations around the world.)

Once the attention of the device manufacturer has been captured, the C&I system manufacturer has to persuade him to divulge the development plans for his product range. The reason for this is simple: the time-scale involved in developing and launching a new computer-based system may be two or three years. This time-scale is comparable with the entire life expectancy of a single generation of microprocessors. In other words, if the C&I systems manufacturer launches a development programme based on a variety of microprocessor that is available today, the device will be obsolete by the time the system is launched.

The C&I manufacturer therefore has to start designing his system and building prototypes using early development samples of microprocessors and so forth. At this stage his designs must be based on a great deal of predicted data, some of which may prove to be overoptimistic.

At the end of all this, the new system is launched and after early trials it becomes widely used (or so the supplier hopes). Because of this success, the manufacturer of a rival system then begins to lose his own market share, so he too starts looking at a new development. This is eventually launched. A complex race, an expensive game of 'technical leap-frog', is then on.

Even without this competition, each new generation of microprocessor offers advances in terms of power and performance, and offers solutions to problems that usage of the previous generation has exposed. Worse still, the manufacturers of the electronic components cannot support endless varieties of devices, so the day eventually arrives when the older generation of components is withdrawn. This adds further complexity to the life of the C&I manufacturer who has a responsibility to keep supplying his customers with replacement parts over several decades.

10.2 Living with change

If the above sounds like an excuse, it is, but it is a very valid excuse, because the pressures are real, and they are inexorable. The only realistic approach therefore, is to accept that change is inevitable and to formulate a plan to deal with it. All the same, when a system is purchased every possible precaution should be taken to ensure that it provides satisfactory service over a long period. Some precautions are listed below:

10.2.1 Defining the requirements for long-term support

As might be expected, much depends on the actions that are taken at the outset, even before the system has been ordered. At the definition stage of a DCS contract an important target is to wring from the system vendor as many written assurances of long-term support as possible, and to make as sure as is possible that the vendor is reliable and that he will be able to stand by his promises. It is a hard goal to meet, but the time to press for such promises is when an order for several hundreds of thousands of pounds, sometimes millions, is in the offing.

Unfortunately, the commercial clauses in large contracts are usually framed by people who know little or nothing about the evolution of DCS or C&I technology and the effects on plant maintenance. In this situation a vendor can respond with weasel words that are worth very little in reality.

Having obtained the promises, it is vital that they are examined as carefully as possible by professionals who really understand the relevant technology. The time and cost of this evaluation will be repaid many times, over the operational life of the plant.

10.2.2 Keeping track of costs

Once the equipment has been acquired, the users must focus on how best to use it, and they must be prepared to ignore all information on new systems (which are always better than the one that has just been bought!). The fact is that it has been bought, and now the users must gain the maximum advantage from the investment.

An important procedure is to keep careful records of all the markers along the road that stretches from initial acquisition to ultimate replacement:

- Log each failure that occurs, recording the time and date of the event. This is useful information for the user and the vendor, since it supports or disproves reliability claims.
- Keep a note of the time that elapses between placing an order and receiving the goods. This will help reinforce your arguments that delays in shipping spares are increasing, exposing the plant to increased levels of risk, causing reduced output, or incurring increased operational costs.
- Monitor and record how long it takes to replace the failed item of equipment with a spare from stock.
- Record the cost of spare parts (printed-circuit cards, power supplies etc.).
- Note the delivery time for spares.
- When recording details of a failure, make a note of any plant outage or reduction in output that can be directly attributed to the failure.
- It is also useful to try and relate the failure to any event that occurred at or just before the time at which the failure occurred, such as a severe thunderstorm, or the failure of any other electrical plant or machines. This can assist in post-mortem analyses which may point to a design

defect such as inadequate screening or earthing, or poor design of power supplies.

Table 10.1 shows a simple format that can be used for recording this type of information.

10.3 Making the decision to change

The day will eventually dawn when it is wise to examine the arguments for and against upgrading the DCS, and to then decide on the best course of action to take. Reaching a decision will be eased if careful records have been maintained as described above, but even then pressures will be applied to upgrade the system.

10.3.1 The impact of change on the operators

System refurbishment can produce major changes to the operator interfaces. Such changes are greatest when a system based on the use of a hard-wired desk is replaced with a so-called 'soft desk'.

In the former, every actuator is controlled via its own discrete hand/auto station (alternatively called an auto/manual station). These enabled the operator to completely disconnect the electronic controller from the actuator and to manually position the final element via a discrete control circuit. Such a system provided a high degree of redundancy since, with the exception of the wires linking the control desk to the actuator, the circuit for providing manual control was totally separate from the electronics of the automatic controller.

Hard desks also provided discrete indicators (often moving-coil meters) showing the important plant parameters. Sometimes these indicators were driven by transmitters which were completely separate from those feeding the electronic controller, again providing a high degree of redundancy.

With the advent of high-reliability systems arguments began to be raised that the cost of providing such levels of redundancy was no longer justified. But it was only with the advent of sophisticated fault-detection systems, coupled with highly reliable control computers (often embodying two-out-of-three fault-detection systems) that the more conservative engineers began to accept that it was safe to entrust the control of large-scale hazardous processes to a soft-desk concept.

With boiler control systems, even if such a decision is reached, care should be taken to ensure that essential information is available to the operator in the event of a serious failure of the control system or its power-supply system. For example, there is a need to provide density compensation for certain measurements if a DCS failure occurs. (An example is the drum-level measurement, where the requirements for compensation in the presence of a DCS failure were pointed out in Section 6.3.)

10.4 A refurbishment case study

Changing to a 'soft' desk results in the operator being presented with a system of displays and controls which is completely different from that with which he or she was familiar when using the hard-wired predecessor.

It may be useful to briefly examine a particular example, the control desk and console of a large coal-fired boiler. The original control-room facilities had been supplied in the days of analogue electronic systems (*c.* 1975) and had vertical panels providing full indication and control facilities of all eight mills, plus a multiplexing system that enabled the operator to gain access from the desk to any two of the mills at a given time. At the time of the original design, this presented some difficulties for ensuring that gaining control of a pair of mills did not introduce any disturbance into the plant.

This system was replaced by a distributed system in 1989 (Figure 10.1) but this was, in turn, upgraded with more up-to-date systems in 1998. Figure 10.2 shows the new 'soft-desk' control-room facilities after the second refurbishment.

The use of screen-based control-room facilities enables a wealth of information to be provided for the operator in a form that is easy to assimilate and use. A 'soft desk' is also inherently more flexible, allowing additional control and display facilities to be easily added as requirements evolve or new plant is added.

10.4.1 The involvement of plant staff

Experience has shown that the success of 'soft desk' upgrades can be correlated to the level of ownership taken by the unit operators. The current strategy of many of the DCS & SCADA implementers is to involve the operators in the process of designing the mimic displays, so that they are happy to use the new screens and thus take ownership in the success of the scheme.

In any case, dramatic changes such as these necessitate a programme of retraining for all plant personnel. The operators must learn how to use the new display and control facilities and the maintenance staff must be

Figure 10.1 *The 'hard-wired' control console and panel of a large coal-fired power-plant before refurbishment*

Figure 10.2 *A 'soft desk' replacement for the 'hard-wired' operator facilities after refurbishment*

trained in the maintenance of the new systems. It is of course essential that the costs of such training are considered in the cost/benefit analysis of a refurbishment project.

10.5 Why refurbish?

Plant managers will be constantly besieged by DCS sales personnel who will be only too happy to provide plenty of arguments to justify replacement of the existing systems with their own. To counterbalance this weight of evidence, here are a couple of examples of reasons why a system should not be upgraded.

10.5.1 'The performance of the present system has degraded'

An operational control system that has at one time met its performance requirements is unlikely to start producing inferior results merely because of deterioration of the electronic components. A component failure may occasionally result in drift, or it may cause the system to operate wrongly, but the solution in such cases is to replace the component or the card in which it is fitted.

Deterioration of a DCS's ability to control the plant is far more likely to be the result of poor performance of the interface transducers (the transmitters and actuators), rather than any factor in the system itself. For example, inadequate maintenance of a valve actuator will result in hysteresis or overshoot. Changing the electronic system that controls the valve will do nothing to improve on control performance.

Equally likely is the possibility that the controlled element itself (a valve or damper) requires maintenance. In several cases, the application of grease to a shaft or bearing has worked wonders on the performance of the loop!

The point is this: if the performance of an actuator or a final element has deteriorated because of lack of maintenance, it is unlikely that a major investment in a new DCS will result in any dramatic long-term improvement. It was the attitudes of the management or staff that resulted in the lack of maintenance and these will not necessarily change when the new system arrives. A short-term improvement may well be obtained at the outset (because the manufacturer will have been forced to fine-tune the performance of all actuators and final elements in order to meet performance guarantees, and because any new system engenders some enthusiasm among maintenance staff, if not the plant operators), but once the situation has settled down the lack of maintenance will once again result in an eventual degradation in performance.

10.5.2 New systems are faster, more powerful, better

Users will always be aware of a system that seems to work better in some way than the one installed on their own plant. That's the nature of the game.

But the system installed on any installation has been adjusted to work effectively with the idiosyncrasies of the plant. It will have taken somebody a long time to reach that stage. If the system is changed that learning curve will have to be renegotiated when the system is replaced. This underlines the importance of documentation, as described in the following section.

10.6 Documenting the present system configuration

It is almost inevitable that the design of a control system will alter between the initial conceptual stage, through handover and during prolonged use afterwards. The initial system configuration will be carefully defined before commissioning begins. Although the subsequent changes may be recorded, all too often the reasons for introducing them may not be logged. Although the changes may be the result of errors, it is equally likely that they are required because of some misunderstanding, during the design phase, of the plant's characteristics or functions.

The changes are implemented in order to correct such errors or misunderstandings, so that the system works better. But if the reasons for the changes are forgotten, there is every likelihood that when the time comes to refurbish or replace the systems the original misunderstandings will be repeated.

When embarking on a replacement or refurbishment project, the task will be immeasurably assisted by the availability of comprehensive and accurate documentation of the systems as finally configured, together with detailed information on the reasons behind all changes to the original designs.

10.7 Summary

We have now reached the end of this overview of a wide and complex subject. I hope that I have been able to throw some light on the technology and that the explanations may have lifted some of the veils of mystery that sometimes seem to obscure it. The fact remains that it is a complex matter and it is unwise to entrust the safety of a plant to people who do not understand either the control aspects or the plant operations.

It has been my privilege to have worked with power plant throughout my career and I hope that some of what I have put down here will be useful, and that it may encourage others to take up a very interesting and important subject.

Table 10.1 A form for recording failures and their possible causes, and related factors

Log No.	Failed item		Date of failure	Time of failure	Nature of failure (off calibration, total loss, etc.)	Effects of failure (loss of generation, increased manning levels, cost of additional fuel, resources etc.)	Failure of other plant or machines immediately before or at the same time	Other factors (electrical storms, exceptional heat or humidity etc.)	Replacement			Signatures
	Part No.	Serial No.							Date ordered	Order No.	Date received	
A	A	A	A	A	A	B	C	D	E	F	G	
1	B1288	384723	05 July 98	14:24	Card ouput lost completely	Unit load reduced to 79 MW for 10 hrs	Equipment-room HVAC plant failed 8 hours earlier	Hot day 32°C	06 Jul 98	SP/09/ 65660	12/8/98	A, B: J Harris C-E: Schmidt G: D Bergoff

Further reading

Boiler and turbine design and construction

BLOCH, H. P.: 'A practical guide to steam turbine technology' (McGraw-Hill, London, 1996)
BS 1113: 1998: 'Specification for design and manufacture of water-tube steam generating plant (including superheaters, reheaters and steel tube economizers', BSI, London, 1998
ELLIOTT, T. C., CHEN, K., and SWANEKAMP, R. C.: 'Standard handbook of powerplant engineering' (McGraw-Hill, London, 1998)
JERVIS, M. W. (Ed.): 'Modern power station practice (BEI/Pergamon, Oxford, 1991)
KEARTON, W. J.: 'Steam turbine theory and practice (Isaac Pitman and Sons, London, 1960)
KEHLHOFFER, R., BACHMANN, R., NIELSON, H., and WARNER, J: 'Combined-cycle gas and steam power plants (Pennwell, Tulsa, OK, USA, 1999)
LEYZEROVICH, A.: 'Large power steam turbines: design and operation' (Pennwell, Tulsa, OK, USA, 1997)
SINGER, J. G. (Ed.): 'Combustion — fossil power.
(ABB/Combustion Engineering Inc., Connecticut, USA, 1991, 4th edn.)

Control and instrumentation

ASTROM, K., and HAGGLUND, T.: 'PID controllers: theory, design and tuning' (American Technical Publishers, Hitchin, UK, 1995, 2nd edn.)
BISSELL, C. C.: 'Control engineering' (Chapman and Hall, London, 1994)
BOLTON, W.: 'Newnes control engineering pocket book' (Newnes, Oxford, 1998)
CONSIDINE, D. M.: 'Process instruments and control handbook' (McGraw-Hill, New York, 1985)
DUKELOW, S. G.: 'The control of boilers (American Technical Publishers, Hitchin, UK, 1991, 2nd edn.)

GILLUM, D. R.: 'Industrial pressure, level and density measurement' (American Technical Publishers, Hitchin, UK, 1995)
HUBER, J. C.: 'Industrial fiber optic networks' (ISA, Research Triangle Park, NC, USA, 1995)
Instrument Society of America. 'Instruments in the power industry' (American Technical Publishers, Hitchin, UK 1990)
Instrument Society of America 'The ISA fieldbus guide (ISA, Research Triangle Park, NC, USA, 1997)
JERVIS, M. W. (Ed.): 'Modern power station practice' (BEI/Pergamon, Oxford, 1991)
JERVIS, M. W.: 'Power station instrumentation' (Institute of Measurement and Control, Butterworth-Heinemann, Oxford, 1993)
JORDAN, J. R.: 'Serial networked field instrumentation' (American Technical Publishers, Hitchin, UK, 1995)
LEVINE, W. S.: 'Control handbook' (CRC/IEEE, Boca Raton, FL, USA, 1996)
LINDSLEY, D. M.: 'Boiler control systems' (McGraw-Hill, London, 1991)
MORRISON, R.: 'Instrumentation fundamentals and applications' Wiley, Chichester, UK, 1984)
NOLTINGK, B. E. (Ed.): 'Instrumentation reference book' (Butterworths, London, 1988)
OLSSON, G., and PIANI, G.: 'Computer systems for automation and control' (Prentice Hall, London, 1992)
PLATT G.: 'Process control: a primer for the non-specialist and newcomer' (American Technical Publishers, Hitchin, UK, 1998, 2nd edn.)
SPITZER, D. W. (Ed.): 'Flow measurement' (American Technical Publishers, Hitchin, UK, 1991)

Human factors (ergonomics) and control-room design

IVERGARD, T. 'Handbook of control room design and ergonomics' (Taylor and Francis, London, 1989)
McCORMICK, E. J., and SANDERS, M. S. 'Human factors in engineering and design' (McGraw-Hill, Durham, NC, USA, 1983)
PHEASANT, S. T.: 'Ergonomics – standards and guidelines for designers' (Taylor and Francis, London, 1986)
SALVENDY, G. (Ed.): 'Handbook of human factors' (Wiley, Chichester, UK)
WAGNER, E.: 'The computer display designer's handbook' (Student Literattur, Chartwell-Bratt, Sweden, 1989)

Interference, electromagnetic compatibility and shielding

Council Directive EMC/89/336/EEC 'The EMC directive'. Council of the European Communities, Brussels, 1989

MORRISON, R. and LEWIS, W. H.: 'Grounding and shielding in facilities' (Wiley, Chichester, UK, 1990)
SCOTT, J. S., and VAN ZYL, C.: Introduction to EMC (Newnes, Oxford, 1997)

Safety and reliability

CHEMICAL INDUSTRIES ASSOCATION: 'A guide to hazard and operability studies'. Chemical Industry Safety and Health Council of the Chemical Industries Association, London
COX, S., and TAIT, R.: 'Safety, reliability and risk management: an integrated approach,' (Butterworths, Oxford, 1998)
GOBLE, W. M.: 'Control system safety evaluation and reliability' (American Technical Publishers, Hitchin, UK, 1998, 2nd edn.)
HEALTH AND SAFETY EXECUTIVE: 'Programmable electronic systems in safety-related applications. HMSO, London
INSTITUTE OF ELECTRICAL AND ELECTRONICS ENGINEERS: Human factors in power plants (IEEE, New York, 1997)
BS EN 60801-2: 1993. 'Electromagnetic compatibility for industrial process measurement and control equipment – Part 2: Electrostatic discharge requirements'. British Standards Institution, London
'Draft Standard 00-58: a guide for HAZOP studies on systems which include a programmable electronic system'. Ministry of Defence Directorate of Standardization, Glasgow, 1995
NFPA 8502-95 'Standard for the prevention of furnace explosions/implosions in multiple burner boilers.' National Fire Protection Association, Quincy, MA, USA
REDMILL, F., and RAJAN, J.: 'Human factors in safety-critical systems' (Butterworth-Heinemann, Oxford, 1997)

Tuning of control loops

ASTROM, K., and HAGGLUND, T.: 'PID controllers: theory, design and tuning' (American Technical Publishers, Hitchin, UK, 1995, 2nd edn.)
CORRIPIO, A. B.: 'Tuning of industrial control systems' (American Technical Publishers, Hitchin, UK, 1990)
McMILLAN, G. K.: 'Tuning and control loop performance: a practitioner's guide, (American Technical Publishers Ltd., Hitchin, UK, 1994, 3rd edn.)

Valves

BAUMANN, H. D.: 'Control valve primer' (American Technical Publishers Ltd., Hitchin, UK, 1998, 3rd edn.)
FISHER CONTROLS COMPANY: 'Control valve handbook' (Fisher Controls Company, Rochester, Kent. 1977)

INSTRUMENT SOCIETY OF AMERICA: 'ISA handbook of control valves' (American Technical Publishers, Hitchin, UK, 1976, 2nd edn.)

General

FRIEDMAN, P. G.: 'Economics of control improvement' (American Technical Publishers, Hitchin, UK, 1994)
ISA: 'Dictionary of measurement and control' (American Technical Publishers Ltd., Hitchin, UK, 1995, 3rd edn.)
ISA: 'Instrumentation standards and practices (American Technical Publishers, Hitchin, UK, 1991, 10th edn.)
BATTIKHA, N.E.: 'The management of control systems: justification and technical auditing' (American Technical Publishers, Hitchin, UK, 1992).

Index